CESSNA 206
Training Manual

By
Oleg Roud and Danielle Bruckert

Published by Red Sky Ventures, Memel CATS
Copyright © 2010

Contact the Authors:

D Bruckert
redskyventures@gmail.com
+264 81 244 6336
PO Box 11288 Windhoek, Namibia
Red Sky Ventures

O Roud
roudoleg@yahoo.com
+264 81 208 0566
PO Box 30421 Windhoek, Namibia
Memel CATS

Published By Red Sky Ventures and Memel CATS
First edition 2010, this edition March 2011.
Copyright © Oleg Roud and Danielle Bruckert

Createspace
ISBN 13 digit 978-1456376505
ISBN 10 digit 1456376500
Lulu Paperback
ISBN 978-0-557-75281-2

COPYRIGHT & DISCLAIMER

All rights reserved. No part of this manual may be reproduced for commercial use in any form or by any means without the prior written permission of the authors.

This Training Manual is intended to supplement information received from your flight instructor and approved flight training organisation. It should be used for training purposes only, not for operational use in flight, and is not part of the Civil Aviation Authority or FAA approved Aircraft Operating Manual or Pilot's Operating Handbook. While every effort has been made to ensure completeness and accuracy, the approved aircraft flight manual or pilot's operating handbook should be used as final reference. The authors cannot accept responsibility of any kind from the misuse of this material.

ACKNOWLEDGEMENTS:

Peter Hartmann, Aviation Centre Pty Ltd, Windhoek: Provision of technical information, access to maintenance manuals and CD's for authors' research.
Mack Air, Maun, Botswana: Assistance with operational information, and review of first draft.
Brenda Whittaker, Christchurch, New Zealand: Editor, Non Technical.

Note-
ENGLISH SPELLING has been used in this text, which differs slightly from that used by Cessna. Differences in spelling have no bearing on interpretation.

FACTS AT A GLANCE

Common Name: Cessna 206
ICAO Designator: C206
Type: Fixed gear, one to six seat light single engine passenger or utility aircraft. (Note: figures vary between models and serial numbers)

Powerplants	
206G	One 225kW (300hp) Continental IO-520-L fuel injected flat six piston engine driving a three blade constant speed McCauley prop.
206H	One 225kW (300hp) Textron Lycoming IO-540-AC1A driving a three blade constant speed prop.
T206H	One 231kW (310hp) turbocharged TIO-540-AJ1A, Textron Lycoming IO-540-AC1A driving a three blade constant speed prop.
Performance	
206G	Maximum speed at sea level 156kts* Cruise speed 75% power at 6500ft 147kts* Initial rate of climb at sea level 920ft/min Service ceiling 14,800ft Maximum range with reserves standard tanks (59 Gal) 555nm (4.8hours) Takeoff run sea level 275m (900ft), total distance to 50ft obstacle 545 (1780ft). *Performance figures are from POH, which may be a little optimistic.
206H	Maximum speed at sea level 150kts (278km/h); Cruising speed at 75% power at 6500ft 143kts (265km/h); Initial rate of climb 920ft/min; Service ceiling 16,000ft; Takeoff distance sea level 275m (900ft), total distance to 50ft obstacle 570m (1860ft).
T206H	Max speed 315km/h (170kt); Cruising speed at 75% at 20,000ft 165kts (306km/h); Initial rate of climb 1010ft/min; Service ceiling 27,000ft; Takeoff distance 255m (835ft), total distance to 50ft obstacle 535m (1750ft).
Weights	
206G	Stationair II (6 seats) Standard empty weight 898kgs (1977lbs) Utility Version (1 seat) Standard empty weight 852kgs (1875lbs), Maximum ramp 1642kgs (3612lbs), Maximum takeoff and landing 1636kg (3600lb)
206H	Standard empty weight 974kg (2146lb), maximum ramp weight 1640kg (3614lb), Maximum takeoff and landing 1636kg (3600lb),

T206H	Standard empty weight 1011kg (2227lb), maximum ramp weight 1641kg (3616lb), Maximum takeoff and landing 1636kg (3600lb)
Dimensions	
206G	Wing span 10.97m (36ft), length 8.62m (28ft 3in), maximum height (9ft 7.5in)
206H T206H	Wing span 10.92m (35ft 10in), length 8.62m (28ft 3in), height 2.92m (9ft 7in). Wing area 16.2m2 (174sq ft).
Capacity	
C205 and C206's all have a typical seating for six adults, with the C206 utility option coming standard with only one seat. The C207 seats seven or eight.	

QRG/Block Operating Information (MAUW unless specified)

Takeoff:	Short field speed at 50ft 65kts; Normal takeoff speed at 50ft 70-80kts.
Climb:	Vy 84kts Sea Level 78kts 10,000ft, Vx 66kts Sea Level, 70kts 10,000ft; Normal Climb 95-105kts or 500fpm.
Block Cruise:	120kts at 60lt/hr, 5000-7500ft AGL
Landing:	Short field 65kts flap 40; Normal 75-85kts flap up, 65-75kts flap 40.
Emergency:	EFATO 80kts flaps up, 70kts flaps down; Glide 75kts at MAUW, reduce 5kts/400lbs.

Table of Contents

Introduction..10
    History..11
    Cessna 205..11
    Cessna 206..12
        Cessna U206..12
        Cessna P206..12
        Cessna 206H..13
    Cessna 207..13
    Models Differences Table..14
    Modifications..17
        Common Modification's Table..18
Terminology 19
Factors and Formulas..23
    Conversion Factors...23
    Formulas...24
Pilot's Operating Handbook..25
AIRCRAFT TECHNICAL INFORMATION.....................................26
General...26
Airframe...28
    Seats and Seat Adjustment..30
    Doors 31
        Door Handles..31
        Cabin and Door Dimensions..34
        Operation Without the Cargo Door....................................35
        Flap Interrupt Switch 35
        Evacuation Considerations..36
        Windows..36
        Baggage Compartment 36
Flight Controls..38
    Elevator..38
    Ailerons..38
        Differential and Frise Design...39
    Rudder..39
        Stowable Rudder Pedals 40
    Trim...40
        Electric Trim...41
    Flaps...42
        Electric Flap 42
        Note on Use of Flap...43
        Toggle Switch 43
        Flap on Robertson STOL Conversion...............................45
Landing Gear..46
    Shock Absorption..46
    Brakes...47
        Park Brake 48

Towing	49
Engine	50
Engine Profile Diagrams	51
Engine Data Tables	52
Engine General Description	53
Engine Controls	54
Throttle	54
Manifold Pressure and Throttle Setting	55
Full Throttle Height	55
Pitch Control	56
Propeller Governor	56
Summary of High/Low RPM Function	56
Propeller Governor Schematic	57
Propeller Pitch Control	57
Mixture	58
Mixture Setting	58
Throttle Quadrant	59
Engine Gauges	60
Manifold Pressure Gauge	60
Fuel Flow Gauge	61
Tachometer	61
Pressure and Temperature Gauges	62
CHT Gauge	63
EGT Indicator	63
Turbocharged Engines	64
Turbo System Schematic	65
Induction System	66
Oil System	67
Ignition System	69
Dead Cut and Live Mag Check	69
Cooling System	71
Oil Cooler	71
Operation of Cowl Flaps	72
Other Cooling Methods	72
Fuel System	74
Fuel Tanks	74
Fuel System Schematic	75
Bladder Tanks	76
Tip Tanks	77
Fuel Selector and Shut-off Valve	77
Refuelling	78
Filler Cap Quantity	78
Fuel Venting	78
Fuel Drains	79
Fuel Measuring and Indication	80
Auxiliary Fuel Pump and Priming System	81
Priming on Continental versus Lycoming	83

Vapour Locks in the Fuel System	83
Fuel Injection System	83
Fuel Injection System Schematic	84
Electrical System	85
Battery	85
Alternator/Generator	85
Electrical Equipment	86
System Protection and Distribution	86
Electrical System Schematic	90
Flight Instruments and Associated Systems	91
G1000 Data Source Diagram	92
Pitot-Static Instruments	93
Pitot-Static System Diagram - Conventional	94
Pitot-Static System Diagram - Glass	95
Vacuum Operated Gyro Instruments	96
Stall Warning	97
Avionics	98
Audio Selector	98
Intercom	98
VHF Radio Operations	98
Transponder	99
Ancillary Systems	100
Lighting	100
Cabin Heating and Ventilating System	101
Cabin Heating and Ventilating Schematic	102
FLIGHT OPERATIONS	103
NORMAL FLIGHT PROCEDURES	103
Pre-flight Inspection	103
Cabin	104
Exterior Inspection	105
Final Inspection	111
Passenger Briefing	112
Starting	112
Priming, Purging and Flooded Starts	114
Priming	114
Priming Lycoming versus Continental	114
Purging Fuel Vapour	115
Flooded Starts	115
Pre-Heat	116
Starting Procedure	116
Starting the C206G and Earlier models	117
Starting the C206H	117
After Start	118
Warm Up	119
Taxi	119
Engine Run-up	120
Pre-Takeoff Vital Actions	121

CESSNA 206 TRAINING MANUAL

Line-Up Checks..122
Takeoff...122
    Fuel flow Setting for Takeoff..123
    Wing Flap Setting on Takeoff..123
    Normal Takeoff...124
    Short Field Takeoff..124
    Soft Field Takeoff...126
    Crosswind Component..126
    Takeoff Profile..126
    After Takeoff Checks...128
Climb..128
Cruise...130
Descent..131
Approach and Landing 133
    Final Approach Speed...134
    Short Field Landing..135
    Crosswind Landing..135
    Flapless Landing...136
Balked Landing 136
After Landing Checks...136
    Taxi and Shutdown..137
Circuit Pattern..138
Note on Checks and Checklists..142
    Do-Lists...143
Flight Operating Tips..143
    Loading..144
    Systems Management...144
    Engine Handling..144
        Application of Power...145
        Changes of Power...145
        Power During Descents..146
        Mixture Changes..146
        Use of Cowl Flaps..147
        Fuel and Engine Monitoring...147
        Extreme Hot and Extreme Cold Weather Operations...........................147
    Turbocharged Engine Handling 148
        Over-boosting..148
        Spool Up...148
        Cooling Prior to Shutdown..149
NON NORMAL FLIGHT PROCEDURES...150
    Stalling and Spinning..150
    Electrical Malfunctions..150
        Excessive Rate of Charge...150
        Insufficient Rate Of Charge..151
    Abnormal Oil Pressure and Temperature..151
    Rough Running Engine...152
    Magneto Faults...152

Spark Plug Faults..152
　　Spark Plug Fouling..152
　　Spark Plug Failure...153
Engine Driven Fuel Pump Failure..153
Excessive Fuel Vapour..154
Blocked Intake Filter (with Alternate Air Source)...154
Inadvertent Icing Encounter...154
Static Source Blocked...155
EMERGENCY FLIGHT PROCEDURES..156
　　General..156
　　Emergency During Takeoff 156
　　Engine Failure...156
　　　　Engine Failure after Takeoff (EFATO)...157
　　　　Gliding and Forced Landing...158
　　Engine Fire...160
　　Electrical Fire..161
　　Emergency Exit Procedures – Cargo Version..162
PERFORMANCE SPECIFICATIONS..163
GROUND PLANNING..167
　　Weight and Balance...168
　　Performance Graphs and Worksheets..170
　　　　Takeoff Performance..170
　　　　Climb Performance..171
　　　　Cruise Performance...172
　　　　Landing Distance..174
　　　　Non-manufacturer Performance Factors..175
　　　　Ground Planning Worksheets and In-flight Logs................................177
REVIEW QUESTIONS..182

Introduction

This training manual provides technical and operational descriptions of the Cessna 206 aircraft model range.

Information is provided in the introduction on the model C205 and C207, for background information on the model development.

The technical and operational information contained within the book is provided for the Cessna 206 series only.

The information is intended as an instructional aid to assist with conversion and or ab-initio training in conjunction with an approved training organisation and use of the manufacturer's operating handbook. The text is arranged according to the progression typically followed during training to allow easier use by students and assimilation with training programmes. This layout differs from the Pilot's Operating Handbook, which is laid out for easy operational use.

This material does not supersede, nor is it meant to substitute any of the manufacturer's operation manuals. The material presented has been prepared from the basic design data obtained in the Pilot's Operating Handbook, engineering manuals and from operational experience.

Illustration 1a C206 Utility (Cargo) Version

History

The Cessna aircraft company has a long and rich history. Founder Clyde Cessna built his first aeroplane in 1911, and taught himself to fly it! He went on to build a number of innovative aeroplanes, including several race and award winning designs.

In 1934, Clyde's nephew, Dwane Wallace, fresh out of college, took over as head of the company. During the depression years Dwane acted as everything from floor sweeper to CEO, even personally flying company planes in air races (several of which he won!). Under Wallace's leadership, the Cessna Aircraft Company eventually became the most successful general aviation company of all time.

The Cessna 205, 206, and 207, known variously as the Super Skywagon, Super Skylane and Stationair, are a family of single engine, general aviation aircraft with fixed landing gear and may be used in commercial air service or for personal use. The family was originally developed from the popular retractable-gear Cessna 210.

The Cessna 206 family is best known for the powerful engine, rugged construction, large cabin and loading capacity. These features have made the aircraft popular 'bush planes' and for aerial work such as skydiving or photography, they can also be equipped with amphibious floats and skis. The combined total number of Cessna 205, 206 and 207 produced so far is over 8500.

Cessna 205

In its initial form the 205 (originally 210-5) was essentially a fixed undercarriage derivative of the 210 Centurion. Although designated as a 1963 model the 205 was introduced to the Cessna lineup late in 1962, followed by the C205A in 1964.

The C205 is powered by the same 260hp IO-470 engine as the 210B and featured an additional small cargo door on the left side of the fuselage.

The 205 retained the early 210's engine cowling bulge, originally where the 210 stowed its nose wheel on retraction (the space where the nose wheel would have retracted was used for radio equipment in the 205). This distinctive cowling was made more streamlined on the later Cessna 206. There were only 577 Cessna 205's produced, before being replaced by the popular Cessna 206.

Cessna 206

The six-seat Cessna 206 was introduced as a 1964 model and was built until 1986, when Cessna halted production of its single-engine product. It was then re-introduced in 1998 and remains in production at the time of publication. The total number of Cessna 206's produced is now over 6500.

Unlike the C210, from which it is based, the C206 has had relatively few changes over the years. The main changes include the engine (1964 and 1998), electrical system (1965 and 1973) and maximum weight (1967).

Cessna U206

The original 1964 model was the U206, powered by a 285hp Continental IO-520-A. The "U" designation indicated "utility" and this model was equipped with a pilot side door and two opposing rear doors, permitting more convenient access to the back two rows of seats, and permitting easy loading of over-sized cargo.

The TU206 offered a turbocharged version of the U206, powered by the Continental TSIO-520-C engine producing 285hp. In 1967 the turbo TU206 was powered by a TSIO-520-F providing 300hp. The additional 15hp was available at a higher rpm, but was limited to 5 minutes for takeoff and produced a significant noise penalty.

From 1964 to 1969 the U206 was known as the "Super Skywagon". From 1970 it was named the "Stationair", a contraction of "Station Wagon of the Air", which is a good description of the aircraft's intended role.

In 1977 the U206 had its engine upgraded to a Continental IO-520-F of 300 hp (continuous rating, obtained at a more reasonable rpm speed than the previous IO-520-F) and the TU206 engine was changed to the TSIO-520-M producing 310hp.

Production of all versions of the U206 was halted in 1986 when Cessna stopped manufacturing all piston engine aircraft. A total of 5208 U206's had been produced.

Cessna P206

1965 saw the P206 added to the line. In this case the "P" stood for "people", as the P206 had passenger doors on both sides, similar to the Cessna 210 from which it originated.

The P206 was produced from 1965 to 1970 and was powered by a Continental IO-520-A of 285hp. There was a turbocharged model designated TP206 which was powered by a Continental TSIO-520-A also of 285hp.

647 P206's were produced under the name "Super Skylane" which incorrectly made it sound like a version of the Cessna 182.

Cessna 206H

After a production break of twelve years, Cessna started manufacturing a new version of the 206 in 1998, with the introduction of the 206H. The "H" model is generally similar to the previous U206 configuration, with a pilot entry door and double rear doors for access to the middle and back seats. The C206H is marketed under the name "Stationair", and Cessna aptly portrays it as the "Sport Utility Vehicle of the air".

The 206H is powered by a Lycoming IO-540-AC1A powerplant producing 300hp. The turbocharged T206H is powered by a Lycoming TSIO-540-AJ1A engine of 310hp.

Both the 206H and the T206H remain in production in 2008. By the end of 2004 Cessna had produced 221 206H's and 505 T206H's, for a total production of 726 "H" models.

Cessna 207

The Model 207 was a seven and later eight seat development of the 206, achieved by stretching the design further to allow space for more seats. The nose section was extended 18" by adding a constant-section nose baggage compartment between the passenger compartment and the engine firewall; the aft section was extended by 44" by inserting a constant-area section in the fuselage area just aft of the aft wing attach point. Thus the propeller's ground clearance was unaffected by the change (the nose wheel had moved forward the same distance as the propeller), but the tail moved aft relative to the main wheel position, which made landing (without striking the tail skid on the runway) a greater challenge. The move gave that aircraft a larger turning radius, since the distance between main wheels and nose wheel increased by 18 inches but the nose wheel's maximum allowed deflection was not increased.

The 207 was introduced as a 1969 model featuring a Continental IO-520-F engine of 300hp. A turbocharged version was equipped with a TSIO-520-G of the same output.

At the beginning of production the model was called a Cessna 207 "Skywagon", but in 1977 the name was changed to "Stationair 7". 1977 also saw a change in engine on the turbocharged version to a Continental TSIO-520-M producing 310hp – the same engine used in the TU206 of the same vintage.

The 207 added a seat in 1980 and was then known as the "Stationair 8". Production of the 207 was completed in 1984, just two years before U206 production halted. A total of 788 Cessna 207's were manufactured.

The Cessna Model 207 has been popular with air taxi companies, particularly on short runs where its full seating capacity could be used. Very few of these aircraft have seen private use.

Models Differences Table

A brief outline of the models by year with major changes is outlined in the table below.

During practical training, reference should be made to the flight manual of the aeroplane you will be flying to ensure that the limitations applicable for that aeroplane are adhered to. Likewise when flying different models it should always be remembered that MAUW, flap limitations, engine limitations and speeds may vary between models and with modifications. Before flying different models, particularly if maximum performance is required, the POH of the aircraft you are flying should be reviewed to verify differences.

TYPE	NAME	YEAR	MODEL	MAJOR DIFFERENCES
C205		1963	205 0001-0480	3300lbs maximum takeoff weight, IO470 engine; essentially a C210B with fixed gear and electric flap
C205A		1964	205 0481-0577	
C206	Super Skywagon	1964	206 0001-0275	Engine changed to IO520
U206	Super Skywagon (Utility Cargo Door)	1965	206 0276-0437	First cargo door version, 14V Alternator replaces Generator
P206	Super Skywagon (Passenger Door)	1965	P206 0001-0160	First C206 to come out with 6 seats as a standard (not optional) fitting
P206	Super Skylane	1965		

TYPE	NAME	YEAR	MODEL	MAJOR DIFFERENCES
U206A	Super Skywagon (Utility Cargo Door)	1966	U206 0438-0656	Maximum takeoff weight increased to 3600lbs
U206B		1967	U206 0657-0914	
U206C		1968	U206 0915	
TU206A	Turbo-System Super Skywagon (Utility Cargo Door)	1966	U206 0438-0656	
TU206B		1967	U206 0657-0914	
TU206		1968	U206 0915	
P206A	Super Skylane	1966	P206 0161-0306	
TP206A	Turbo-System Super Skylane	1966	P206 0161-0306	
P206A	Super Skylane	1966	P206 0161-0306	
P206B		1967	P206 0307-0419	
P206C		1968	P206-0420	
TP206A	Turbo-System Super Skylane	1966	P206 0161-0306	
TP206B		1967	P206 0307-0419	
TP206C		1968	P206-0420	
TU206D U206D	Super Skywagon Turbo-System Super Skywagon	1969	U206-1235 U206-1444	
P206D TP206D	Super Skylane Turbo-System Super Skylane	1969	P206-0520 P206-0603	
U206E	Super Skywagon	1970	U20601445-U20601587	
TU206E	Turbo-System Super Skywagon	1970		
P206E	Super Skylane	1970	P20600604-647	
TP206E	Turbo-System Super Skylane	1970	P20600604-647	
U206E	Stationair Turbo Stationair	1971	U20601588-1700	
U206F	Stationair Turbo Stationair	1972	U20601701-1874	Flap toggle switch changed to pre-select lever
		1973	U20601875-2199	12V battery changed to 24V

TYPE	NAME	YEAR	MODEL	MAJOR DIFFERENCES
U206F	Stationair Turbo Stationair	1974	U20602200-2579	
U206G	Stationair Turbo Stationair	1975-76	U20602580-3021	
	Stationair II Turbo Stationair II	1977	U20603522-4074	
U206G	Stationair 6 Turbo Stationair 6 Stationair 6 II Turbo Stationair 6 II	1978 1979 1980 1981 1982 1983 1984	U20604075-4649 U20604650-5309 U20605310-5919 U20605920-6439 U20606440-6699 U20606700-6788 U20606789-6846	In 1979 the bladder tanks were changed to integral wet wing tanks.
U206G	Stationair Turbo Stationair Stationair With Value Group A Turbo Stationair II With Value Group A	1985 1986	U20606847-6920 U20606921-7020	
207	Skywagon 207	1969-1977	20700001-0414	New model C207 introduced, 3800lbs Gross weight, 300 or 310hp IO520 series engine Wing span 432-439", length 381" 7 or 8 place seating
T207	Turbo Skywagon 207	1969-1977	20700415-0562	
207	Stationair 7 Turbo Stationair 7 Stationair 7 II Turbo Stationair 7 II	1978-1979	20700563-0788	
C207	Stationair 8 Turbo Stationair 8 Stationair 8 II Turbo Stationair 8 II	1980-1984		
206H	Stationair	1998 On	20608001 on	Lycoming IO/TIO-540 engine, annunciator systems. From 2005, G1000 glass avionics optional, from 2007 standard.
T206H	Stationair TC	1998 On	T20608001 on	

Modifications

Common modifications include the famous cargo pod, floats, most of the common STOL kits (eg. Robertson and Sportsman), additional fuel tanks and various engine modifications including a turbine version. Details on common modifications available are outlined in the table on the following page.

At present there is no 'RG' (retractable gear) version of the C206, as offered with the 100 series Cessnas This is presumably because of the similarity and success of the retractable C210 on which the C206 was based.

Illustration 1b C206 with Cargo Pod

Illustration 1c C206 on Floats

Common Modification's Table

TYPE	NAME and MANUFACTURER	DIFFERENCES and FEATURES
Any	Cargo Pod	(Various) Extra cargo/luggage room, small speed penalty
Any	Skis / Floats	(Various)
Any	Soloy	Turbine Engine Installation, 418 SHP Allison C20S engine
Any	Engine Conversion, Bonaire	Conversion to IO550 engine, 300hp maximum continuous
Any	Engine Conversion, Atlantic Aero	Conversion to IO550 engine, 300hp maximum continuous
Any	Low Fuel Warning System, O & N Aircraft Modifications	Warns when fuel remaining is less than approximately 7USG
F, G, H	Engine Conversion, Thielert	300 or 310hp V8 diesel engine installation
Any	Fuel Cap Monarch Air	Umbrella style fuel caps which fix problems with leaks, predominantly occurring in older flush mounted caps, (available for most Cessna types)
Any	Wing Tip Tanks, Flint Aero	Two auxiliary tip tanks of 16.5USG in each, used with an electrical transfer pump to each main tank. Higher MTOW (3800lbs) is permitted if tanks are half full. Wing length is also increased by 26 inches.
Any	Horton STOL	Tip and wing surface modifications to permit lower stall speed, take-off and landing speeds and distances
Any	Robertson STOL	Increased lift, more speed, added stability, and lower stall speed, take-off and landing speeds and distances. ?

Note: The table above is included for interest and awareness, as there are many C206s operating with the modifications installed, some modifications may no longer be available for installation.

Terminology

Airspeed		
KIAS	Knots Indicated Airspeed	Speed in knots as indicated on the airspeed indicator.
KCAS	Knots Calibrated Airspeed	KIAS corrected for instrument error. Note this error is often negligible and CAS may be omitted from calculations.
KTAS	Knots True Airspeed	KCAS corrected for density (altitude and temperature) error.
Va	Max Manoeuvering Speed	The maximum speed for full or abrupt control inputs.
Vfe	Maximum Flap Extended Speed	The highest speed permitted with flap extended. Indicated by the top of the white arc.
Vno	Maximum Structural Cruising Speed	Sometimes referred to as "normal operating range". Should not be exceeded except in smooth conditions and only with caution. Indicated by the green arc.
Vne	Never Exceed speed	Maximum speed permitted, exceeding will cause structural damage. Indicated by the upper red line.
Vs	Stall Speed	The minimum speed before loss of control in the normal cruise configuration. Indicated by the bottom of the green arc. Sometimes referred to as minimum 'steady flight' speed.
Vso	Stall Speed Landing Configuration	The minimum speed before loss of control in the landing configuration, at the most forward C of G*. Indicated by the bottom of the white arc.
*forward centre of gravity gives a higher stall speed and so is used for certification		
Vx	Best Angle of Climb Speed	The speed which results in the maximum gain in altitude for a given horizontal distance.
Vy	Best Rate of Climb Speed	The speed which results in the maximum gain in altitude for a given time, indicated by the maximum rate of climb for the conditions on the VSI.
Vref	Reference Speed	The minimum safe approach speed, calculated as 1.3 x Vso.

Vbug	**Nominated Speed**	The speed nominated as indicated by the speed bug, for approach this is Vref plus a safety margin for conditions.
Vr	**Rotation Speed**	The speed which rotation should be initiated.
Vat	**Barrier Speed**	The speed to maintain at the 50ft barrier or on reaching 50ft above the runway.
	Maximum Demonstrated Crosswind	The maximum demonstrated crosswind during testing.
Meteorological Terms		
OAT	**Outside Air Temperature**	Free outside air temperature, or indicated outside air temperature corrected for gauge, position and ram air errors.
IOAT	**Indicated Outside Air Temperature**	Temperature indicated on the temperature gauge.
ISA	**International Standard Atmosphere**	The ICAO international atmosphere, as defined in document 7488. Approximate conditions are a sea level temperature of 15 degrees with a lapse rate of 1.98 degrees per 1000ft, and a sea level pressure of 1013mb with a lapse rate of 1mb per 30ft.
	Standard Temperature	The temperature in the International Standard atmosphere for the associated level, and is 15 degrees Celsius at sea level decreased by two degrees every 1000ft.
	Pressure Altitude	The altitude in the International Standard Atmosphere with a sea level pressure of 1013 and a standard reduction of 1mb per 30ft. Pressure Altitude would be observed with the altimeter subscale set to 1013.
	Density Altitude	The altitude that the prevailing density would occur in the International Standard Atmosphere, and can be found by correcting Pressure Altitude for temperature deviations.
Engine Terms		
BHP	**Brake Horse Power**	The power developed by the engine (actual power available will have some transmission losses).
RPM	**Revolutions per Minute**	Engine drive and propeller speed.

	Static RPM	The maximum RPM obtained during stationery full throttle operation
Weight and Balance Terms		
	Moment Arm	The horizontal distance in inches from reference datum line to the centre of gravity of the item concerned, or from the datum to the item 'station'.
C of G	Centre of Gravity	The point about which an aeroplane would balance if it were possible to suspend it at that point. It is the mass centre of the aeroplane, or the theoretical point at which entire weight of the aeroplane is assumed to be concentrated. It may be expressed in percent of MAC (mean aerodynamic chord) or in inches from the reference datum.
	Centre of Gravity Limit	The specified forward and aft points beyond which the CG must not be located. Typically, the forward limit primarily effects the controllability of aircraft and aft limits stability of the aircraft.
	Datum (reference datum)	An imaginary vertical plane or line from which all measurements of arm are taken. The datum is established by the manufacturer.
	Moment	The product of the weight of an item multiplied by its arm and expressed in inch-pounds. The total moment is the weight of the aeroplane multiplied by distance between the datum and the CG.
MZFW	Maximum Zero Fuel Weight	The maximum permissible weight to prevent exceeding the wing bending limits. This limit is not always applicable for aircraft with small fuel loads.
BEW	Basic Empty Weight	The weight of an empty aeroplane, including permanently installed equipment, fixed ballast, full oil and unusable fuel, and is that specified on the aircraft mass and balance documentation for each individual aircraft.
SEW	Standard Empty Weight	The basic empty weight of a standard aeroplane, specified in the POH, and is an average weight given for performance considerations and calculations.
OEW	Operating Empty Weight	The weight of the aircraft with crew, unusable fuel, and operational items (galley etc.).

	Payload	The weight the aircraft can carry with the pilot and fuel on board.
MRW	**Maximum Ramp Weight**	The maximum weight for ramp manoeuvring, the maximum takeoff weight plus additional fuel for start taxi and runup.
MTOW	**Maximum Takeoff Weight**	The maximum permissible takeoff weight and sometimes called the maximum all up weight, landing weight is normally lower as allows for burn off and carries shock loads on touchdown.
MLW	**Maximum Landing Weight**	Maximum permissible weight for landing. Sometimes this is the same as the takeoff weight for smaller aircraft.

Note: In recent texts there is a trend towards the use of the correct term 'mass' instead of 'weight', effectively replacing the W with M in all the above abbreviations. In everyday language, and in most aircraft manuals and pilot operating handbooks, the term weight (although technically incorrect) remains in common use. For this reason it has been used here. In this context there is no difference in the applied meaning.

Other

AFM	**Aircraft Flight Manual**	These terms are inter-changeable and refer to the approved manufacturer's handbook. General Aviation manufacturers from 1976 began using the term 'Pilot's Operating Handbook', early manuals were called Owner's Manual and most legal texts use the term AFM.
POH	**Pilot's Operating Handbook**	
PIM	**Pilot Information Manual**	A Pilot Information Manual is a new term, coined to refer to a POH or AFM which is not issued to a specific aircraft.

Factors and Formulas

Conversion Factors

lbs to kg	1kg = 2.204lbs	kgs to lbs	1lb = .454kgs
USG to lt	1USG = 3.785Lt	lt to USG	1lt = 0.264USG
lt to Imp Gal	1lt = 0.22 Imp G	Imp.Gal to lt	1Imp G = 4.55lt
nm to km	1nm = 1.852km	km to nm	1km = 0.54nm
nm to St.m to ft	1nm = 1.15stm 1nm = 6080ft	St.m to nm to ft	1 st.m = 0.87nm 1 st.m =5280ft
feet to meters	1 FT = 0.3048 m	meters to feet	1 m = 3.281 FT
inches to cm	1 inch = 2.54cm	cm to inches	1cm = 0.394"
Hpa(mb) to "Hg	1mb = .029536"	" Hg to Hpa (mb)	1" = 33.8mb

AVGAS FUEL Volume / weight SG = 0.72

Litres	Lt/kg	kgs	Litres	lbs/lts	Lbs
1.39	1	0.72	0.631	1	1.58

Crosswind component per 10 kts of wind

Kts	10	20	30	40	50	60	70	80
10	2	3	5	6	8	9	9	10

Formulas

Celsius (C) to Fahrenheit (F)	C = 5/9 x(F-32), F = Cx9/5+32
Pressure altitude (PA)	PA = Altitude AMSL + 30 x (1013-QNH)
Standard Temperature (ST)	ST = 15 – 2 x PA/1000 ie. 2 degrees cooler per 1000ft altitude
Density altitude (DA)	DA = PA +(-) 120ft/deg above (below) ST i.e. 120Ft higher for every degree hotter than standard
Specific Gravity	SG x volume in litres = weight in kgs
One in 60 rule	1 degree of arc $\cong$ 1nm at a radius of 60nm i.e degrees of arc approximately equal length of arc at a radius of 60nm
Rate 1 Turn Radius	R = GS/60/π, $\cong$ GS/20
Rate 1 Turn Bank Angle	Degrees of Bank $\cong$ G/S/10+7
Percent to fpm	fpm $\cong$ % x G/S Or fpm = % x G/S x 1.013
Percent to Degrees	TANGENT (radians) x100 = Gradient in % INVERSE TANGENT (%/100) = Angle in Radians
Degrees to Radians	Degrees x π / 180 = radians
Gust factor (Rule of Thumb)	Vbug = Vref+1/2HWC + Gust eg. Wind 20kts gusting 25 at 30 degrees to Runway: Vbug = Vref +.7x10+5 = Vref+12, If the Vref is 75kts, Vat should be 75+12 = 87kts

Pilot's Operating Handbook

The approved manufacturer's operating handbook, is issued for the specific model and serial number, and includes all applicable supplements and modifications done on that aircraft. It is legally required to be on board the aircraft during flight, and is the master document for all flight information.

In 1975, the US General Aviation Manufacturer's Association introduced the 'GAMA Specification No. 1' format for light aircraft manufacturer's handbook, calling it a 'Pilot's Operating Handbook' (POH). This format was later adopted by ICAO in their Guidance Document 9516, and is now required for all newly certified light aircraft by ICAO member states. Most light aircraft listed as built in 1976 or later, have a POH in this format.

This format was designed for the best ergonomic use during flight. It is recommended that pilots become familiar with the order and contents of each section, as summarised in the table below.

Section 1	General	Definitions and abbreviations
Section 2	Limitations	Specific operating limits, placards and specifications
Section 3	Emergencies	Complete descriptions of action in the event of any emergency or non-normal situation
Section 4	Normal Operations	Complete descriptions of required actions for all normal situations
Section 5	Performance	Performance graphs for takeoff, climb, cruise, and landing, and often includes stall speeds, CAS and crosswind calculation
Section 6	Weight and Balance	Loading specifications, limitations and loading graphs or tables
Section 7	Systems Descriptions	Technical descriptions of the entire aircraft, engine and all systems
Section 8	Servicing and maintenance	Maintenance requirements, inspections, storage, oil requirements, towing and handling.
Section 9	Supplements	Supplement sections follow the format above for additional equipment or modifications.
Section 10	Safety Information	General safety information and helpful operational recommendations

For use in training this text should be read in conjunction with the POH from on board the aircraft you are going to be flying. Even if you have a copy of a POH for the same model C206, the aircraft you are flying may have supplements for modifications and optional equipment which affect the operational performance.

AIRCRAFT TECHNICAL INFORMATION

General

The Cessna 206 aircraft is a single-engine, high-wing monoplane of an all metal, semi-monocoque construction. Wings are externally braced with a single strut attached to the fuselage and contain sealed sections forming integral or bladder type fuel bays.

The fixed tricycle landing gear consists of tubular spring-steel main gear struts and a steerable nose wheel with an air-hydraulic fluid shock strut.

The six place seating arrangement is of conventional, forward facing type.

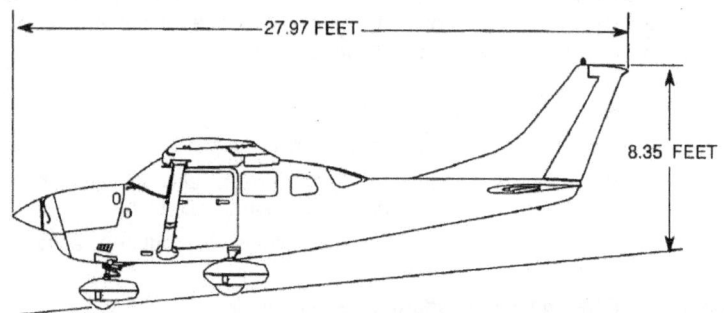

Illustration 2a C206H Left Profile

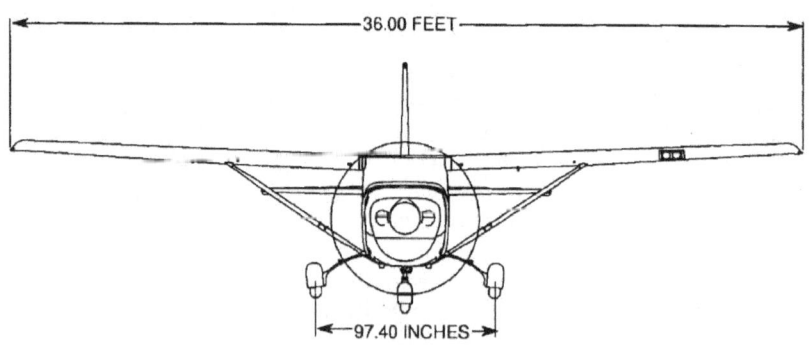

Illustration 2b C206H Front Profile

The standard power plant installation is a horizontally-opposed, air-cooled, six-cylinder, fuel injected engine driving an all-metal, constant-speed propeller. The engine is typically normally aspirated, however higher performance is offered in the turbocharged version of the Model 206.

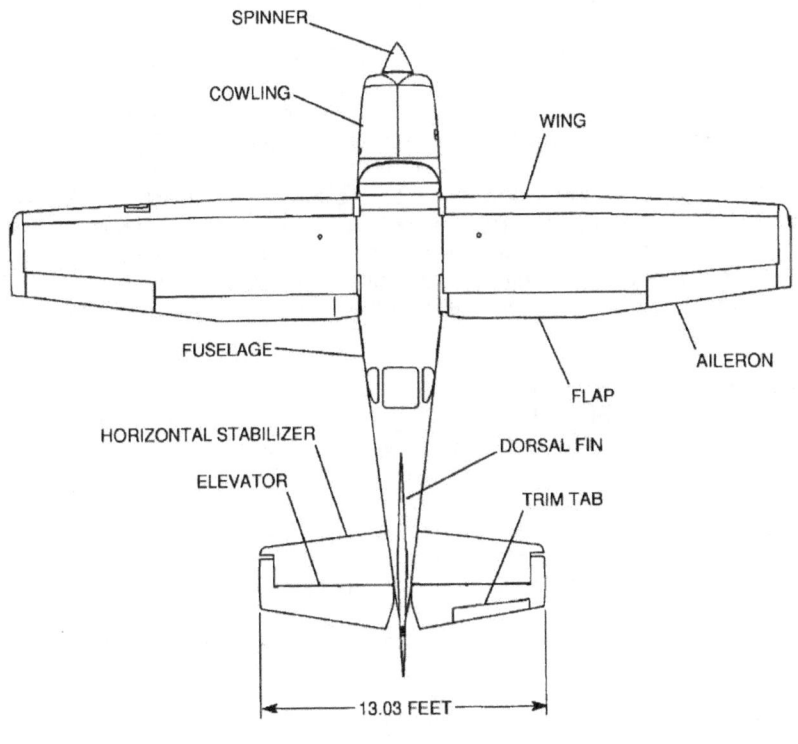

Illustration 2c C206H Top View

Airframe

The airframe is a conventional design similar to other Cessna aircraft you may have flown (for example the C152, C172).

The construction is a semi-monocoque type consisting of formed sheet metal bulkheads, stringers and stressed skin.

Semi-monocoque construction is a light framework covered by skin that carries much of the stress. It is a combination of the best features of a strut-type structure, in which the internal framework carries almost all of the stress, and the pure monocoque where all stress is carried by the skin.

The fuselage forms the main body of the aircraft to which the wings, tail section and undercarriage are attached. The main structural features are:
+ front and rear carry through spars for wing attachment;
+ a bulkhead and forgings for landing gear attachment;
+ four stringers for engine mounting attached to the forward door posts.

Each all-metal wing panel is a full cantilever type, with a single main spar, two fuel spars, formed ribs and stringers. The front fuel spar also serves as an auxiliary spar and provides the forward attachment point for the wing. An inboard section of the wing, forward of the main spar, is sealed to form an integral (i.e. non-bladder) fuel bay area. Stressed skin is riveted to the spars, ribs and stringers to complete the structure. An all-metal, balanced aileron, flap, and a detachable wing tip are part of
each wing assembly. A navigation light is mounted in each wing tip.

The empennage or tail assembly consists of the vertical stabilizer and rudder, horizontal stabilizer and elevator.

The fin is primarily of metal construction, consisting of ribs and spars covered with an aluminium skin. Fin tips are glass fibre or plastic composite construction. Hinge brackets at the rear spar attach the rudder.

The horizontal stabilizer is primarily of metal construction, consisting of ribs and a front and rear spar which extends throughout the full span of the stabilizer. The skin is riveted to both spars and ribs. Stabilizer tips are constructed of plastic composite. The elevator tab actuator screw is contained within the horizontal stabilizer assembly, and is supported by a bracket riveted to the rear spar. The underside of the stabilizer contains an opening which provides access to the elevator tab actuator screw. Hinges on the rear spar support the elevator.

The construction of the wing and empennage sections consists of:
- a front (vertical stabilizer) or front and rear spar (wings/horizontal stabilizer)
- formed sheet metal ribs
- doublers and stringers
- wrap around and formed sheet metal/aluminium skin panels
- fibreglass tips
- control surfaces, flap and trim assembly and associated linkages

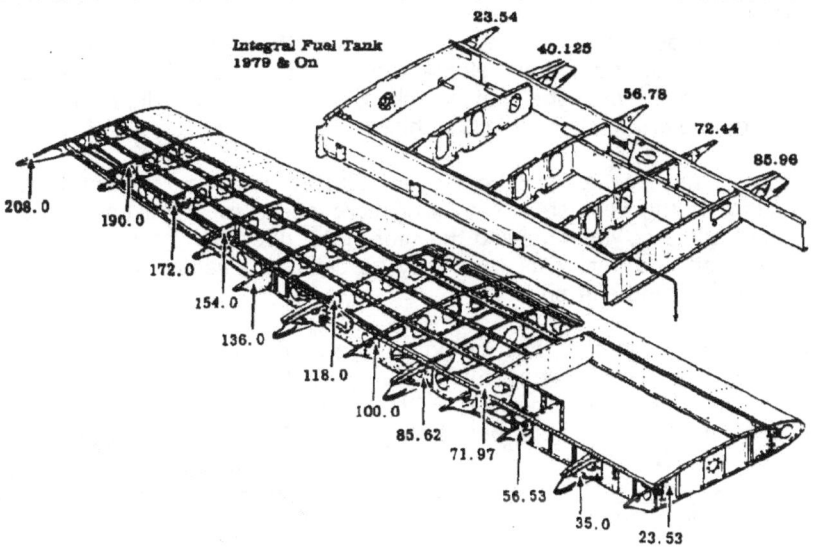

Illustration 2d Wing Cross Section

Seats and Seat Adjustment

The seating arrangement consists of:
+ Cargo/Utility Versions: One seat only (more seats may be optionally fitted)
+ Passenger Versions: Six individual seats

The pilot and copilot seats are adjustable in forward and aft position, and normally also for both seat height and back inclination.

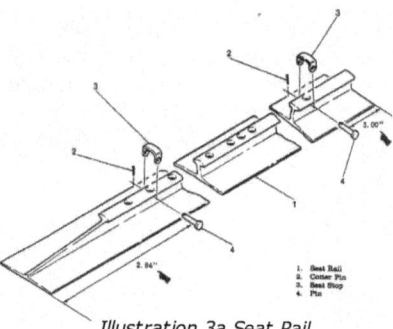

Illustration 3a Seat Rail

Position the seat forward or aft by lifting the tubular handle under the centre of the seat bottom, and sliding the seat into position, then release the handle and check that the seat is locked in place (when the seat is locked correctly the handle pins fit down fully into one of the aligned pairs of holes in the seat rail).

Seat rails have a stop at the forward and aft limits to prevent the seat derailing. If the seat stop is missing care needs to be taken to prevent derailing the seat when moving the seat in the direction of the missing stop. A small bolt and nut or lock pin may be used as a temporary stop until a replacement can be fitted.

Raise or lower the seat by rotating a large crank under the right corner of the seat. Seat back angle is adjustable by rotating a small crank under the left corner of the seat. The seat bottom angle will change as the seat back angle changes, providing proper support.

Seat back and height should be adjusted to ensure adequate visibility and control before start-up.

When adjusting the seats forward and aft care should be taken to ensure the position is locked. Additional seat locks are available, and have been installed on many aircraft following accidents involving failure or improper positioning of the primary seat lock resulting in slipping of seat position during critical phases of flight. In the aircraft with the seat lock installed, to move the pilot seat backward the seat lock shall be held in open position while moving the seat.

The seat backs may be folded full forward providing easy access to the passengers' seats.
All seats are removable by removing the seat stop and sliding the seat forwards or rearwards until the seat is free from the rails.

Doors

The more commonly produced cargo or utility versions (U206 and 206H models) have one door on the left side of the cockpit and two doors, in a clamshell configuration, on the right rear side of the fuselage.

Illustration 3b U206 Cargo Door

The P206 has the traditional configuration of two front doors on the left and right side and a small cargo door at the rear, similar to the C210 and other types in the Cessna single engine series.

Door Handles

Both the passenger and the forward part of the cargo interior doors are locked by rotating the handle 90 degrees from the open position.

To open the door from outside the aircraft, utilize the recessed door handle near the aft edge of each door. Depress the forward end of the handle to rotate it out of its recess, and then pull outward. To close or open the doors from inside the aircraft, use the inside door handle in the arm rest.

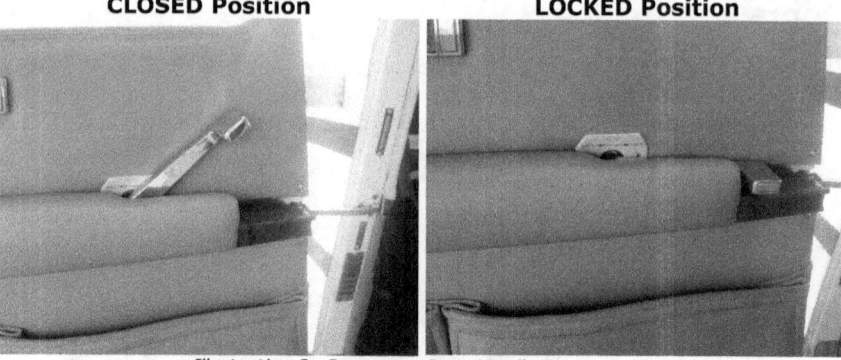

Illustration 3c Passenger Door Handle Positions

Most handles have three positions OPEN, CLOSE, and LOCK. The handle is spring-loaded to the OPEN (up) position. When the door has been pulled shut and latched, it can be locked by rotating the door handle forward to the LOCKED position (flush with the arm rest).

Illustration 3d Cargo Door Handle and Latch

Door Latches

The door latch, on both the forward and rear doors, is similar on most single engine Cessnas, consisting of a small rack and pinion type latch. The pinion being a round gear on the airframe, and the rack a flat gear on the door. When the teeth become warn it may become difficult to mesh the locking mechanism without pressure on the door. It is also possible to achieve locking only on the last tooth of the rack, where upon vibration or forces in flight may cause the door to open.

The security of the door should be checked by applying positive pressure prior to takeoff.

Door Latch Rack   **Door Latch Pinion**

Illustration 3e Door Latching Mechanism

Handle modifications are available with an operating pin that ensures the door is in the correct position when closed, preventing the handle from being lowered if the pin is not flush with the door recess. These modifications are recommended and minimize the risks of doors inadvertently opening in flight.

Should the door need to be opened during flight, or if the door is not latched securely, the pressure from the slipstream and relative airflow will make it nearly impossible to latch again. If this occurs it may be required to secure the door with either a shoulder harness or unused seatbelt until a landing can be made. If the flight conditions permit, Cessna recommends to set up the aeroplane in a trimmed condition at approximately 85kts, momentarily shove the door outward slightly, and forcefully close and lock the door.

Cargo Door Latch

The aft rear door has a latch at the top and bottom of the door and a red locking lever on the forward edge of the door. The aft rear door must be closed before the forward rear door can be latched. The front door then latches and operates in the same way described above.

Both cabin doors must be locked prior to flight, and should not be opened intentionally during flight.

Illustration 3f Cargo Door Latch

Cabin and Door Dimensions

The following diagram illustrates the approximate dimensions of the cabin and positions of the cabin doors.

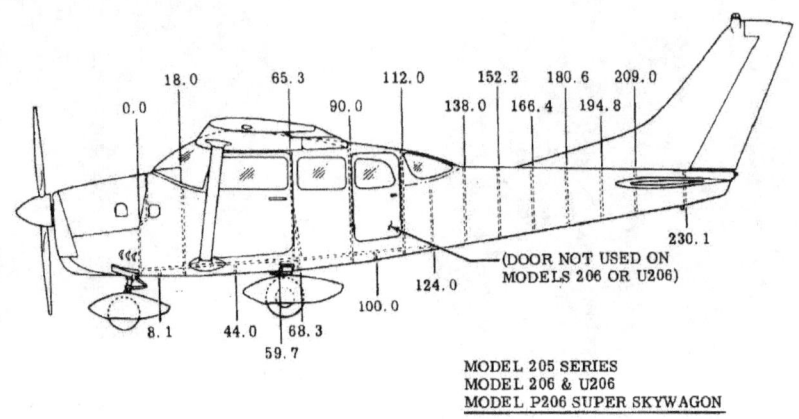

MODEL 205 SERIES
MODEL 206 & U206
MODEL P206 SUPER SKYWAGON

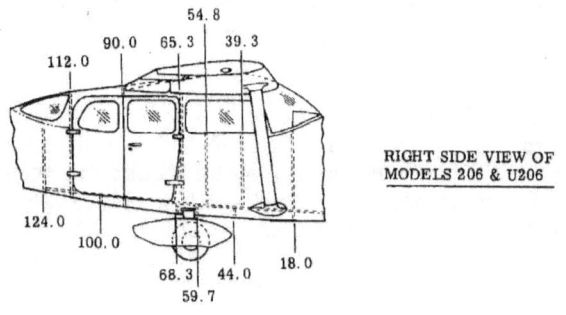

RIGHT SIDE VIEW OF
MODELS 206 & U206

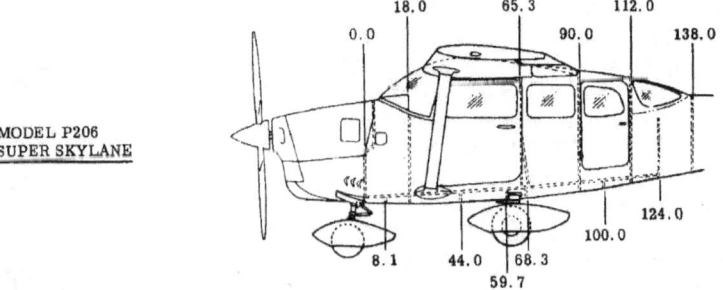

MODEL P206
SUPER SKYLANE

Illustration 3g Typical Door Configurations

Operation Without the Cargo Door

The utility configuration C206 may be operated without the cargo door installed. This is often desirable for photographic or skydive work. If the aircraft is operated without the cargo door a spoiler kit must be fitted. The kit deflects airflow away from the cargo door opening, and permits flap operation (see flap interrupt switch in following section).

Flap Interrupt Switch

The rear cargo door, on the utility configuration of the C206, is positioned directly inboard of the flaps. This means that should the flaps be lowered while the door is open damage will occur to either the door or the flaps or both.

✻ Models prior to the Serial No. 206-0196, the front cargo door must be either in full open or full closed position before operating wing flaps, or damage to the flaps will result. Extreme care should be taken when selecting flap on these models, in particular during preflight or ground operations with passengers who may open the rear door for air.

Illustration 3h Flap over Cargo Door and Flap Interrupt Switch

All other models with aft cargo doors contain a flap interrupt switch, mounted on the front cargo door frame, which prevents flap operation while the front cargo door is open. The flap interrupt switch can be fitted on the models which were not fitted with a flap interrupt switch. This modification is highly recommended to avoid accidental damage.

If an aircraft is operated with the cargo doors removed, a 'spoiler kit' must be installed. The spoiler kit contains a switch depressor for the flap interrupt switch to permit operation of the flap.

A faulty or stuck micro-switch or incorrectly latched door can be a common cause for flap failure. Inspection of the door and switch may be an easy way to resolve this fault.

Evacuation Considerations

When operating a C206 equipped with rear cargo doors, it is very important to remember the consequences of the flap position, and have a thorough emergency escape plan, to provide for the event of flaps being jammed or left down after an emergency landing or ditching. Many unnecessary fatalities have resulted after ditching accidents where the occupants could not escape through the one remaining forward exit.

To open the rear door with the flaps extended the passenger must perform the complicated procedure of opening the front part as far as possible (about 2 inches) then open the rear part of the door and re-stow the rear door handle. With the handle stowed, there is enough clearance to open the rear part of the door, whilst the front part remains blocked shut by the flap.

The cargo door must have a placard indicating the emergency operating procedure, as shown in the photograph below.

Illustration 3i Cargo Door Emergency Operation Placard

See more details under Emergency Escape in the Emergency Procedures Section.

Windows

Both forward cabin windows can be opened. The window is closed by a latch on the lower edge of the window frame. The latch has a groove to keep it in the closed position, which is normally released by a small push button on the latch. The window is equipped with a spring-loaded retaining arm which will help rotate the window outward and hold it there.

Illustration 3j Window Latches

Baggage Compartment

The baggage compartment consists of the area from the back of the rear passenger seats to the aft cabin bulkhead. A baggage shelf, above the wheel well, extends aft from the aft cabin bulkhead. Access to the baggage compartment and the shelf is gained through a lockable baggage door on the left side of the aeroplane on passenger versions, or from the double cargo doors on the right side in cargo versions.

When loading the aircraft, children should not be placed or permitted in the baggage compartment. Any material that may be hazardous to the aeroplane or occupants should never be carried anywhere in the aeroplane including the baggage compartment.

During normal passenger operations, the cargo compartment may not be fitted with any protection to separate the cargo compartment and the passengers.
If not properly secured, the baggage can slide forward into the passenger cabin during turbulence or manoeuvring.
A baggage net with six tie-down straps can be purchased for securing baggage, and is attached by tying the straps to tie-down rings provided in the aircraft. This net must be used when operating the aircraft with cargo, and is recommended for all flight situations.

The baggage door (passenger version) has a push latch incorporating a key lock. Due to the low pressure around the aircraft, baggage doors sometimes open during flight and care should be taken to ensure the door is closed securely. For this reason it is recommended whenever possible that the lock is used.

Cargo Pod
The cargo pod, which is often symbolic of the 'Sports Utility Vehicle of the skies' image Cessna promotes, adds significant loading room and prevents the need to load dirty items such as tools, or even skis, in the cabin.
The cargo pod takes approximately 300lbs, and care must be taken not to overload the aircraft with the increased volumetric capacity. The cargo pod has an added benefit of significantly reducing the problem of loading beyond the aft centre of gravity limit, and should normally be loaded first if operating with full passenger seats.
The pod creates more profile drag and so comes with a small speed penalty.

Illustration 3k C206 with Cargo Pod

Flight Controls

The aeroplane's flight control system consists of conventional aileron, rudder and elevator control surfaces. The control surfaces are manually operated through mechanical linkages to the control wheel for the ailerons and elevator, and rudder/brake pedals for the rudder. A manually-operated elevator trim tab is provided and installed on the right elevator.

The control surfaces are formed in a similar way to the wing and tail section with the inclusion of the balance weights, actuation system (control cables etc) and trim tabs. Control actuation is provided by a series of push-pull rods, bellcranks, pulleys and cables with their respective connections.

Elevator

The elevator is hinged to the rear part of the horizontal stabilizer on both sides.

The main features are:
- an inset hinge with balance weights
- an adjustable trim tab on right hand side of the elevator

Illustration 4a Elevator

The elevator is operated by fore-and-aft movement of either the pilot or copilot's control wheel. The elevator control cables at their ends, are attached to a bellcrank mounted on a bulkhead in the tailcone. A push-pull tube connects this bellcrank to the elevator arm assembly, installed between the left and right sides of the elevator.

The outer leading edge of the each elevator, incorporates an extension, which protruding into the airflow, forms an aerodynamic hinge tab. The extension also contains a balance weight, which mechanically assists with control input. Both act to reduce the force required to move the elevator, making the aircraft easier to control.

Ailerons

Conventional hinged ailerons are attached to the trailing edge of the wings. Main features of the aileron design include:
- a forward spar containing aerodynamic "anti-flutter" balance weights;
- "V" type corrugated aluminium skin joined together at the trailing edge;
- differential and Frise design;

Differential and Frise Design

The ailerons incorporate both Differential and Frise design. Differential refers to the larger angle of travel in the up position to the down position, increasing drag on the down-going wing. Frise-Type ailerons are constructed so that the forward part of the up-going aileron protrudes into the air stream below the wing to increase the drag on the down-going wing. Both features acting to reduce the effect of Adverse Aileron Yaw, reducing the required rudder input during balanced turns. These features have the additional advantage of assisting with aerodynamic balancing of the ailerons.

Illustration 4b Ailerons

The left aileron is equipped with a trim tab on the inboard end of the trailing edge, which is adjustable only on the ground.

Both ailerons contain balance weights on the leading edge of the control surface.

Rudder

The rudder forms the aft part of the vertical stabilizer. The main features include
✈ A horn balance tab and balance weight;
✈ An adjustable rudder trim.

The top of the rudder incorporates a leading edge extension which contains a balance weight and aerodynamically assists with control input in the same way as the elevator hinge point.
There is no rudder trim tab on the rudder, trimming is accomplished by trimming whole rudder (discussed in details in later chapter).

Illustration 4c Rudder

Rudder control is maintained through use of conventional rudder pedals which also control nose wheel steering. The system is comprised of the rudder pedal assembly, cables and pulleys, all of which link the pedals to the rudder and nose wheel steering. The rudder movement is limited by stop at 23 degrees either side of neutral.

The rudder pedals are also connected to the nose wheel steering, this is described more fully in the section on landing gear.

The later models are equipped with an aileron-rudder interconnection to provide improved stability in the flight. Moving control column to the right, for example, will move right aileron up and move rudder to the right. This reduces adverse aileron drag, and provides for more coordinated turns with minimum rudder input. The system can be easily overridden if necessary when cross control is required.

Stowable Rudder Pedals

When dual controls are installed, optional stowable rudder pedals are available for the copilot's position to prevent accidental interference by the right seat passenger.

To stow the rudder pedals depress the "rudder-stow" push button and pull the control knob out, this will release the rudder pedals which can be then pressed fully forward against the fire wall. Release the push button leaving the control in the fully out position, where the pedals will be held in position against the firewall by spring clips. To release the pedals from their stowed position, push the control knob fully in, and pull forward on each stowed rudder pedal until they un-latch from their stowed position.

Trim

Both the elevator and rudder contain a pilot adjustable trim tab for balancing aerodynamic forces.

The elevator trim tab is provided on the right side of the elevator, spanning most of the rear section of the right elevator. The trim tab operates conventionally, moving opposite to the control surface, reducing the aerodynamic force on the control surface in order to hold the selected position.

Illustration 4d Trim Wheels

The rudder trim tab provides a means of balancing slipstream forces during changes of power and speed.

Rudder trimming is accomplished through a bungee connected to the rudder control system altering the neutral position of the rudder, and a trim control wheel for actuation mounted on the control pedestal.

Please note, that the rudder control system, rudder trim control system, and the nosewheel steering system are interconnected and adjustments of any one system will affect the others.

Trimming is accomplished through mechanical linkages to the vertically and horizontally mounted trim control wheel as follows:

- ELEVATOR trim: Forward (up) rotation of the trim wheel will trim nose-down, conversely, aft (down) rotation will trim nose-up.

- RUDDER trim: Left rotation of the trim wheel will apply left yaw, right rotation will apply right yaw. That is - Move the trim tab in the direction of the applied rudder until the force required on the rudder pedal to maintain balance is removed.

Electric Trim

An optional electric elevator trim may be installed, generally in conjunction with an autopilot or wing levelling system. The electric elevator trim control consists of two switches mounted on the pilot's control column and a circuit breaker.

To apply the electric elevator trim, both switches on the control column need to be pressed. Engaging of the electric elevator trim can be monitored by the movement of the mechanical trim wheel. The split control provides protection against inadvertent selections, and from trim runaway.

Whenever an electric trim is installed it is important to check the ability to override and to ascertain where the disconnect switch and circuit breaker are located for an urgent disconnect in case of a trim runaway.

Flaps

The flaps are constructed basically the same as the ailerons with the exception of the balance weights and the addition of a formed sheet metal leading edge section.

The maximum deflection of the flaps on all C206 models is 40 degrees. Flap actuation, on all models, is achieved electrically through a flap switch and electric motor.

The wing flaps are of the single-slot, fowler type. Both design features act to further reduce the stalling speed. The single slot modifies the direction of the airflow to maintain laminar flow longer. The fowler design increases the size of the wing surface area on extension.

Electric Flap

The electric flap control system is comprised of:
- → An electrical motor;
- → A transmission assembly;
- → A drive pulley, cables, and push-pull rods;
- → Follow-up control.

Power from the motor and transmission assembly is transmitted to the flaps by the drive pulley, cables and push-pull rods.
Electrical power to the motor is controlled by two microswitches mounted on a floating arm assembly, through a camming lever and follow-up control.
The flaps are extended or retracted by positioning the flap lever on the instrument panel to the desired flap deflection position.

Illustration 4e Electric Flap Lever

The flap lever is moved up or down in a slot in the instrument panel that provides mechanical stops at the 10° and 20° positions. For settings greater than 20°, move the switch level to the right to clear the stop and select the desired position.

A scale and pointer on the left side of the switch level indicates flap travel in degrees. The flap limiting speeds are marked horizontally with coloured tabs on the left of the level.

The flap system is protected by a 15-ampere circuit breaker, labelled FLAP, on the right side of the instrument panel.

When the flap control lever is moved to the desired flap setting, an attached cam trips one of the microswitches, activating the flap motor. As the flaps move to the position selected, the floating arm is rotated by the follow-up control until the active microswitch clears the cam, breaking the circuits and stopping the motor. To reverse flap direction the control lever is moved in the opposite direction causing the cam to trip a second microswitch which reverses the flap motor. The follow-up control moves the cam until it is clear of the second switch, shutting off the flap motor. Failure of a microswitch will render the system inoperative without indication as to why.

Limit switches on flap actuator assembly prevent over-travel of the flaps in the full UP or DOWN positions. Failure of limit switches will cause the motor to continue to run after the desired position is reached.

Note on Use of Flap

The POH specifies 0 to 20 degrees of flaps for normal takeoffs, and 20 degrees for maximum performance takeoffs.

For normal approaches use of full flap generally provides better handling in the flare, smoother transition to landing attitude and speed, lower touch down speeds and better pitching moments.

When landing in strong crosswind conditions with little experience on the aircraft, it is recommended to land with lower flap setting for example 20°. Lower flap settings will improve the lateral stability and the higher touchdown speed will provide better control against the drift (it should be remembered that lower flap setting will require higher V_{ref}).

Full flap can be used in moderate and even strong crosswind conditions depending on pilot comfort level, however full flap crosswind landings should be practised dual, and attempted solo only when sufficient competence has been achieved.

Toggle Switch

Some early model C206 aeroplanes were fitted with a toggle switch for flap actuation.

The switch is a three position, double-throw switch. Flaps may be selected DOWN by holding the switch in the desired position until the setting required is achieved. The flaps may be selected fully up by selecting the UP position, or to an intermediate selection by selecting up then reselecting the neutral position.

The flap position transmitter is mechanically connected to the right flap drive pulley and electrically transmits position to the flap position indicator located on the instrument panel.

In flight at 100 mph, indicated airspeed, the flaps should take approximately 9 seconds to fully extend and 7 seconds to retract. On the ground with engine running the flaps take approximately 7 seconds to extend or retract.

To select from zero to 10 degrees the toggle switch is moved to the down position for 3-4 seconds while monitoring the flap indicator, and then returned to neutral when the desired position is reached, likewise from 10 degrees to 20 degrees etc.

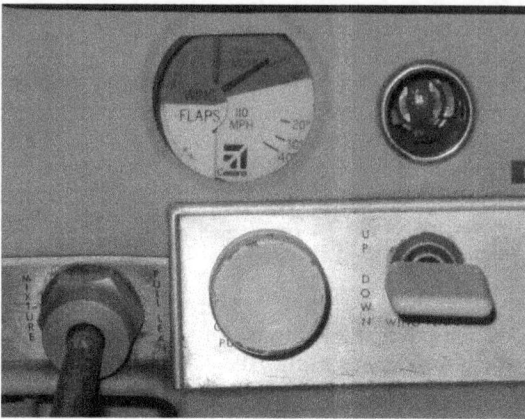

Illustration 4f Flap Toggle Switch

These switches have the inherent design fault of allowing the pilot to easily inadvertently select the flaps to the **fully up or fully down position** if the neutral position is not reselected correctly.

This error invariable occurs in two ways:
- flap was selected up or down and not returned to neutral (ie either the pilot omitted to return the switch to neutral, or the return spring failed), resulting in full travel up or down;
- after selection when returning to neutral the selector is returned too far and instead of neutral the flap begins travelling in the opposite direction.

Should the aircraft you are flying have a toggle switch for a flap lever, remember to take particular care with selection and confirmation of the correct position, to prevent these errors.

A transmission is connected to and actuates the right flap drive pulley. This transmission converts the rotary motion of electric motor to the push-pull motion needed to operate the flaps. The transmission will free-wheel each end of its stroke; therefore, it cannot be damaged by overrunning when lowering or raising the flaps and no adjustments or limit switches are necessary.

Flap on Robertson STOL Conversion

Among other features, the Robertson STOL (Sierra Industries) lowers the ailerons symmetrically when the flaps are extended. In their lowered position, the ailerons increase outboard wing lift by changing the camber of the wing, the same way the flaps do for the inboard wing section.

Note: Different speeds and limitations apply to an aircraft fitted with a Robertson STOL, and it is important to review the RSTOL supplement in the POH for operational requirements.

Landing Gear

The landing gear is of the tricycle type with a steerable nose wheel and two fixed main wheels. The landing gear may be equipped with wheel fairings for reducing drag.

The steerable nose wheel is mounted on a forked bracket attached to an air/oil (oleo) shock strut. The shock strut is secured to the tubular engine mount.

Nose wheel steering is accomplished by two spring-loaded steering bungees linking the nose gear steering collar to the rudder pedal bars. Steering is available up to 10 degrees each side of neutral, after which brakes may be used to gain a maximum deflection of 30 degrees right or left of centre. During flight the nose wheel leg extends fully, bringing a locking mechanism into place which holds the nose wheel central and free from rudder pedal action.

Illustration 5a Nose Wheel

Shock Absorption

Shock absorption on the main gear is provided by the tabular spring-steel main landing gear struts and air/oil nose gear shock strut. Because of this the main gear is far more durable than the nose gear and is thus intended for the full absorption of the landing.

Correct extension of shock strut is important to proper landing gear operation. Too little extension will mean no shock absorption resulting in shock damage during taxi and landing, too much and proper steering will become difficult and premature nose wheel contact on landing may occur. Should the strut extend fully while on the ground the locking mechanism will cause a complete loss of nose wheel steering.

A hydraulic fluid-filled shimmy damper is provided to minimize nose wheel shimmy. The shimmy damper offers resistance to shimmy (nose wheel oscillation) by forcing hydraulic fluid through small orifices in a piston. The dampener piston shaft is secured to a stationary part and the housing is secured to the nose wheel steering collar which moves as the nose wheel is turned right or left, causing relative motion between the dampener shaft and housing. This movement in turn provides the resistance to the small vibrations of the nose wheel.

Illustration 5b Shock Strut and Shimmy Damper

Brakes

Each main gear wheel is equipped with a hydraulically actuated disc-type brake on the inboard side of each wheel.

The hydraulic brake system is comprised of:
- two master cylinders immediately forward of the pilot's rudder pedals
- a brake line and hose connecting each master cylinder to its wheel brake cylinder
- a single-disc, floating cylinder-type brake assembly on each main wheel

The brake master cylinders located immediately forward of the pilot's rudder pedals, are actuated by applying pressure at the top of the rudder pedals. A small reservoir is incorporated into each master cylinder for the fluid supply. The co-pilot (instructor) pedals operate the brakes by a mechanical linkage to the pilot's pedals.

Through the system design it is easily possible to inadvertently apply the brakes whilst under power. This increases wear on brakes and increases stopping distances. Prior to applying brakes to stop the aircraft always ensure the throttle is closed.
Care should also be taken during take-off and landing roll to ensure that brakes are not inadvertently applied when attempting to use the nose wheel steering. Prior to beginning the takeoff roll, or before landing, feet should be kept well clear of the tips of the rudder pedals ("heels on the floor").

Park Brake

The park brake system consists of a control lever below the pilot's side instrument panel which is connected by a cable to linkage at the brake master cylinders. At the brake master cylinders, the control operates locking plates which trap pressure in the system after the master cylinder piston rods have been depressed by toe operation of the rudder pedals.

The method of using the parking brake system is:
1. Apply pressure on the top of the rudder pedals;
2. Pull parking brake control to the out position;
3. Rotate the control downwards to the locked position;
4. Release toe pressure.

To release the parking brake apply the reverse procedure, pull the park brake and rotate in the reverse direction then push fully in towards the control panel.

The park brake should be released when securing the aircraft after installing chocks. If the park brake is left on during prolonged parking, temperature rises will cause expansion of brake fluid, which may cause the brakes to lock. If this occurs, mechanical intervention may be required to unlock the brakes.

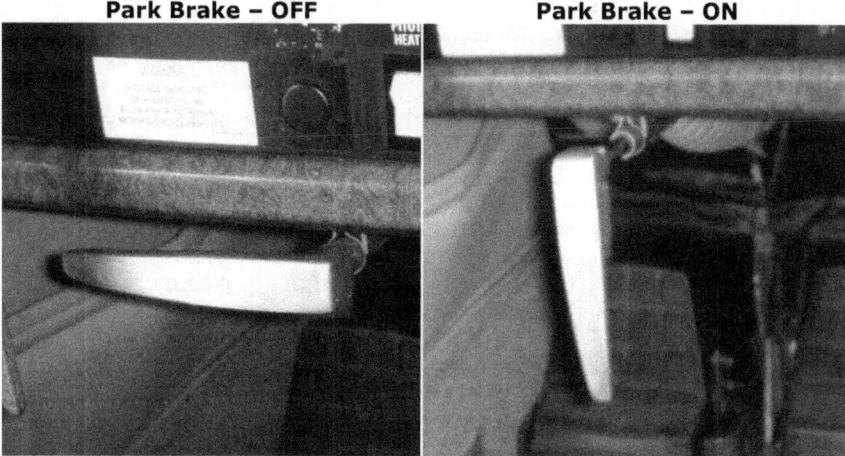

Illustration 5c Park Brake

Towing

Moving the aircraft by hand is best accomplished by using the landing gear struts as a pushing point. A tow bar attached to the nose gear should be used for steering and manoeuvering the aircraft on the ground. When towing the aircraft, never turn the nose wheel more then 30 degrees either side of centre or the nose gear will be damaged.

When no tow bar is available, the aircraft may be manoeuvered by pressing down on the tail section. Do not press on the control surfaces or horizontal/vertical stabilizers as structural damage will occur to the mounting or skin surface. Press down on the tail section or aft fuselage immediately forward of the vertical stabilizer leading edge.

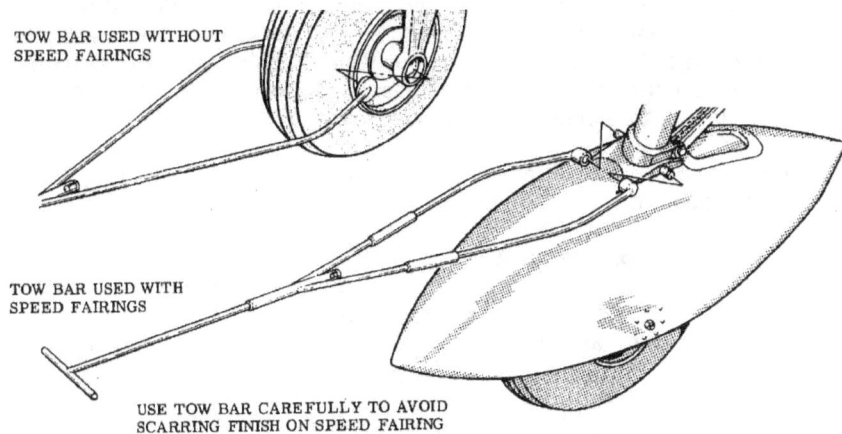

Illustration 5d Towbars

Engine

The aeroplane is powered by a flat 6 cylinder horizontally opposed piston engine.

Some common engine configurations include:
- U206/P206 - One 225kW (300hp) Continental IO-520-L fuel injected, normally aspirated, flat six piston engine driving a three blade constant speed prop.
- T206 - One 230kW (310hp) fuel injected and turbocharged TSIO-520-R, driving a constant speed three blade prop.
- 206H- One 225kW (300hp) Lycoming IO-540 fuel injected, normally aspirated, flat six piston engine driving a three blade constant speed prop.
- T206H - One 230kW (310hp) Lycoming IO-540 fuel injected and turbocharged TSIO-540-R, driving a constant speed three blade prop.
- Bonaire/IO550 Engine Conversion - One 240kW (300hp) Continental IO-550-L fuel injected, normally aspirated, flat six piston engine driving a three blade constant speed prop.

Maximum power may be either maximum continuous or limited to five minutes for takeoff.

In the IO-520 maximum power is 300bhp at 2850rpm, and maximum continuous is 285bhp at 2700rpm. The engine specifications for the IO520 and TSIO520 are included on the following pages. The Bonaire engine develops the maximum 300bhp at 2700rpm with no time limitations at full power.

The propeller is a three bladed, constant speed, aluminium alloy McCauley propeller. The propeller is approximately 2m (80 inches) in diameter. Some models of C206 may be equipped with a three bladed, constant speed, aluminium alloy Hartzel propeller.

Engine Profile Diagrams

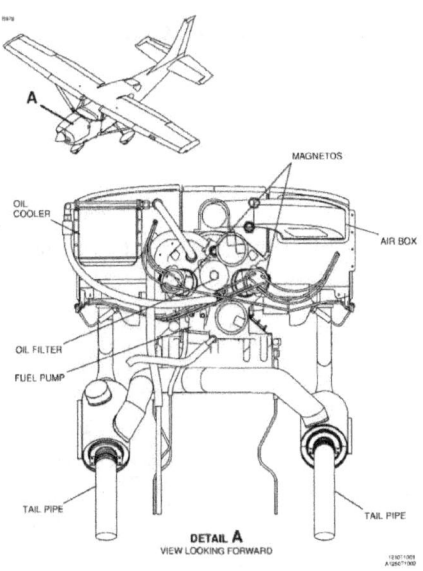

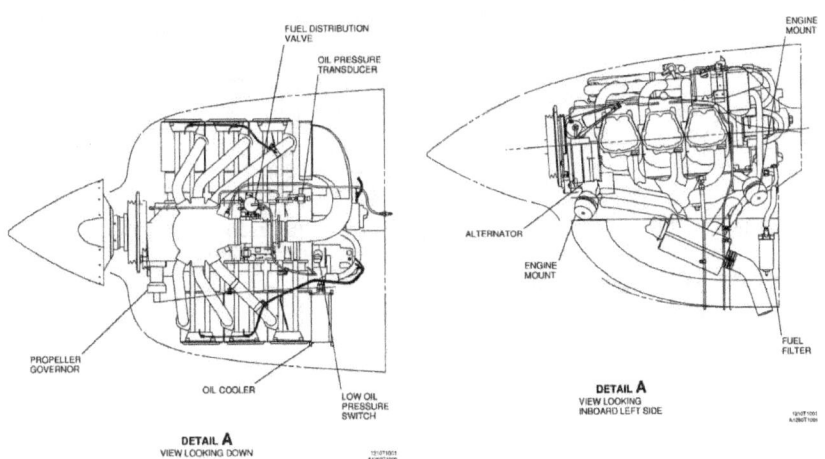

Illustration 6a Engine Profiles

Engine Data Tables

Sample Engine Data IO-540-AC1A5

Table 1. IO-540-AC1A5 Technical Description

Rated Horsepower at 2700 RPM	300
Number of Cylinders	6 Horizontally Opposed
Displacement	541.5 Cubic Inches (7.87 l)
Bore	5.125
Stroke	4.376
Compression Ratio	8.9:1
Firing Order	1-4-5-2-3-6
Magnetos:	
Right Magneto	Slick Model No. 6351 (fires at 20° BTDC)
Left Magneto	Slick Model No. 6351 (fires at 20° BTDC)
Spark Plugs	18MM
Torque	420 Inch-pounds
Valve Rocker Clearance (hydraulic tappets collapsed)	0.028 to 0.080 inch (0.7 to 2.0 mm)
Fuel Injector	PAC RSA-10ED1
Tachometer	Mechanical Drive
Oil Capacity	11.0 Quarts (10.41 l)
Oil Pressure	Minimum 25 PSI Normal 55 to 95 PSI Maximum 115 PSI
Oil Temperature	Normal 165 to 200°F (74 to 93.3°C) Maximum 245°F (118.3°C)
Dry Weight - with accessories	454 Lbs (205.93 kg)

Engine General Description

The Continental IO/TSIO 520 and the Lycoming IO/TIO 540 series engine both have six-cylinders, are horizontally-opposed, air cooled, with a wet-sump, fuel injected system and have a direct drive to a constant-speed propeller. The 'T' Indicates the turbo version respectively.

The cylinders, numbered from rear to front are staggered to permit a separate throw on the crankshaft for each connecting rod. The right rear cylinder is number 1 and cylinders on the right side are identified by odd numbers 1,3 and 5 (with 2,4 and 6 on the left).

Teledyne Continental Motors recommends engine overhaul at approximately 1700 hours operating time depending on the engine model. The Textron Lycoming IO540AC1A5 and TSIO450AJ1A motors are recommended for overhaul at 2000hours (SI1009AS-2006), although some authorities may approve different schedules.

The engine must not be operated above the specified maximum and maximum continuous RPM limitations. Should an inadvertent over speed occur, it must be reported to the aircraft maintenance organisation for further investigation.

Major accessories of the engine include a propeller governor (constant speed drive), dual magnetos, a starter motor, and a belt-driven alternator or generator, a vacuum pump and a full flow oil filter.

The engine fuel system consists of an engine-driven fuel pump and fuel injection system. The fuel injection system is a low pressure system of injecting fuel into the intake valve port of each cylinder. It is a multi-nozzle, continuous-flow type system, which controls fuel flow to match engine airflow. Any change in the throttle position, engine speed, or a combination of both, causes changes in fuel flow in the correct relation to engine airflow. A manual mixture control and a fuel flow indicator are provided for leaning at any combination of altitude and power setting.

The aircraft is equipped with an all-metal, constant-speed propeller. The constant-speed propeller is single-acting, in which engine oil pressure, boosted and regulated by the governor, is used to obtain the correct blade pitch for the engine load.

Engine Controls

The engine controls consist of:
- ✈ Throttle Control
- ✈ Pitch (CSU) Control
- ✈ Mixture Control
- ✈ Cowl Flaps

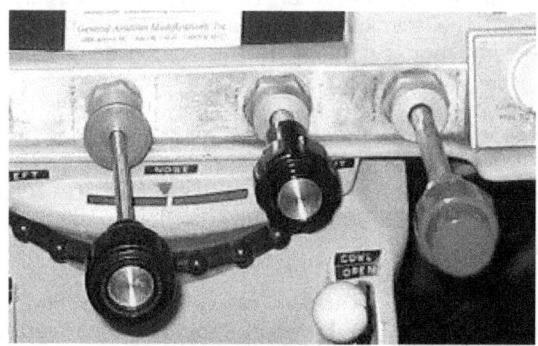

Illustration 6b Engine Controls

The throttle, mixture and propeller controls are of the push-pull type. The propeller and mixture controls are equipped with a vernier lock to fix the desired position and to make fine adjustments.

The cowl flaps are covered fully in the section on engine cooling, however they are included here as it is important that pilots new to operating with cowl flaps learn to treat them as part of the power quadrant.

Throttle

Engine power is controlled by a throttle, located on the lower centre portion of the instrument panel. The throttle control has neither a locking button nor a vernier adjustment like propeller and mixture controls, but contains a knurled friction knob which is rotated for more or less friction as desired. The friction knob acts to prevent vibration induced "creeping" of the control, however throttle position (and resulting manifold pressure RPM) should still be monitored, especially in critical phases of flight.

The throttle provides input to the fuel control unit. The fuel control unit maintains a constant flow of fuel to match the engine airflow using a combination of the throttle setting and the engine speed. Thus for a desired throttle setting and mixture setting at the same density altitude, a certain power setting will result.

Throttle in Open Position	Throttle in Closed Position

Illustration 6c Throttle Butterfly

The throttle control operates in a conventional manner:
- **full forward** position, the throttle is **open** and the engine produces **maximum** power,
- **full aft** position, it is **closed** and the engine is **idling** or windmilling.

Manifold Pressure and Throttle Setting

When the engine is below governing speed the power provided by the throttle is indicated by engine RPM.

The manifold pressure is less than the indicating scale (10 inches), and the propeller is at the fine pitch stop, therefore increases and decreases in engine speed are transmitted directly to the propeller. Throttle changes result in a change in RPM.

Once the engine reaches governing speed, RPM will remain fixed at the selected setting, and the throttle will control the manifold pressure. Engine power is indicated by manifold pressure and the selected RPM is maintained by the Constant Speed Unit (CSU).

Full Throttle Height

Although we are aware of the power reduction with height on a fixed pitch propeller, with a CSU we can see this directly by the manifold pressure - throttle relationship. As we climb and the ambient pressure drops, and to maintain the same manifold pressure (climb power setting) in the climb, the throttle setting must be progressively increased (opened). This will continue up to the point that the throttle is fully forward. This point is termed " full throttle height", climbing above this level will result in reduction of manifold pressure, until we reach the absolute aircraft ceiling where the power is just enough to maintain level flight.

Most normally aspirated aircraft engines will reach full throttle height around 5000ft density altitude.

Pitch Control

The propeller pitch is controlled by the Constant Speed Unit (CSU), which consists of the propeller pitch control knob, propeller governor, and the associated linkages and actuators. The governing function is achieved by altering the propeller blade angle (pitch) to maintain the selected RPM during changes in aircraft speed or power setting.

Propeller Governor

The propeller governor controls flow of engine oil, boosted to high pressure by the governing pump, to or from a piston in the propeller hub.

Engine lubricating oil is supplied to the piston in the propeller hub through the crankshaft.

An increase or decrease in throttle setting or a change in aircraft speed affects the governor balance which maintains the selected RPM. If the throttle setting is increased or if aircraft speed is increased, engine RPM will try to increase. The governor senses this and directs oil pressure to the forward side of the piston. The blades will be moved to a higher pitch and engine speed will remain constant. Conversely, if the throttle opening or the aircraft speed is decreased, the engine RPM will try to decrease RPM, and the governor, sensing this allows oil to drain from the forward side of the piston, spring tension and centrifugal twisting moments will move the blades to a lower pitch to maintain the selected engine speed.

Summary of High/Low RPM Function

In summary:
- Oil pressure acting on the piston turns the blades towards high pitch (low propeller RPM);
- Relief of oil pressure permits the centrifugal force to turn the blades toward low pitch (high RPM).

This system provides redundancy in that loss of oil pressure in the governor will cause the blades to move to the high power position where the system can be operated as a fixed pitch system by reducing the throttle until RPM reduction is achieved.

Complete loss of oil pressure during an engine failure would unfortunately have the same effect.

Propeller Governor Schematic

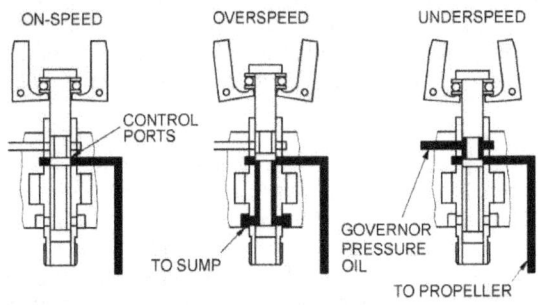

Illustration 6d Propeller Governor Schematic

Propeller Pitch Control

The pitch control knob is the pilot interface of the propeller constant speed unit (CSU). The pitch control, is directly connected to the propeller governor, and is used to set the engine RPM as desired for various flight conditions.

The pitch control is labelled PROP PITCH, PUSH INCR RPM. When the control knob is pushed in, blade pitch angle will decrease, giving a high RPM (fine pitch). Inversely, when the control knob is pulled out, the blade pitch angle increases, thereby decreasing RPM (coarse pitch).

The propeller control knob is equipped with a vernier feature which allows fine (small) RPM adjustment by rotating the knob, clockwise to increase RPM, and counter-clockwise to decrease it. To make rapid adjustment, the button on the end of the control knob must be depressed. With the button depressed, the control can then be positioned freely. The vernier control should be used for all normal flight situations, the override knob is used for ground checks, and for large changes that may be required in emergencies.

The pilot sets the RPM on the pitch control in the cockpit, and providing the power is above the governing range, the prop governor will act to maintain the RPM.

When below the governing range the propeller cannot reduce the pitch angle any more to maintain the desired RPM setting, and rests on the "low pitch stop". This normally occurs at around 10" manifold pressure, where the throttle will continue to control the RPM, and is applicable for most ground operations.

Once the power (throttle) is increased into the governing range, the RPM is controlled by the propeller governor and the setting of the propeller pitch control.

With the pitch control set to maximum and the throttle fully forward the engine must develop the maximum RPM specified (the red line on the RPM indicator). This can be checked in a stationary run-up if needed. Should full RPM not be developed after application of full throttle for take-off, it is an indication that there is a possible fault in the CSU unit, take-off should be discontinued.

Functioning of the CSU is checked during the engine run-up at 1700rpm. The propeller pitch is selected momentarily to coarse and back to full fine, ensuring a RPM drop, manifold pressure increase and oil pressure drop and return. Full fine should be selected to ensure the RPM drop is not more than 300rpm, to avoid excessive loading on the engine. This cycling action should be repeated two to three times, as it also ensures that warm engine oil is cycled through the system providing proper lubrication before full loads are applied.

Mixture

The mixture control, mounted to the right of the throttle, is a red vernier type control. It is used for adjusting fuel/air ratio in the conventional way as follows:

- the **full forward** position is the **fully rich** position (maximum fuel to air ratio);
- the **full aft** position is the **idle cut-off** position (no fuel).

For fine adjustments, the control may be moved forward by rotating the vernier knob clockwise (enriching the mixture), or aft by rotating it counter-clockwise (leaning the mixture). For rapid or large adjustments, the control may be moved forward or aft by depressing the lock button on the end of the control, and then positioning the control as desired. When setting the mixture in flight the vernier should always be used (as with the pitch control).

The mixture control should be set to "full rich" for take-off below 3,000 feet of **density** altitude. Above 3,000 feet, on normally aspirated engines the flight manual requires the mixture to be leaned to the correct setting before take-off.
On a turbo engine the pressure is boosted to be equivalent to that at sea level, therefore the mixture is kept fully rich for take-off.

Mixture Setting

For normally aspirated engines the setting should always be slightly rich of the "peak RPM" or maximum power setting to allow for cooling and prevent detonation. This is achieved by rotating the knob counter-clockwise until maximum RPM is obtained with fixed throttle where upon the RPM begins to decrease on further leaning accompanied by slight rough running as cylinders begin to misfire. Then the control is rotated clockwise until RPM starts to decrease again, normally one turn to reach peak RPM again then two turns thereafter.

For maximum engine power, the mixture should be set, and then checked and adjusted as required during the initial take-off roll to achieve the fuel flow corresponding to the field elevation. The fuel flow required is set according to the placard adjacent to fuel flow indicator. The placard, which is required by the AFM, indicates fuel flow vs altitude for maximum power (2850rpm). To achieve maximum takeoff power, it is essential to set the mixture correctly for elevation. Above 3000ft the power reduction is significant, and this procedure should always be employed for high density altitude airfields.

During enroute climbs a cruising climb at 25 inches of MP, 2550rpm and 100-110 kts, to achieve approximately 500fpm is typically selected to aid engine cooling and passenger comfort. Cruising climb should be conducted at 108 lbs/hr up to 4000 feet and at the fuel flow shown on the Normal Climb Chart in Section 5 for higher altitudes. At approximately 4000 feet the aircraft usually reaches "full throttle height".

The Exhaust Gas Temperature (EGT) Indicator may be used as an aid for mixture leaning in the cruise when operating at 75% power or less. The EGT can also be used as a guide during climb.

To adjust mixture during cruise, lean the mixture to establish the peak EGT, and then enrich the mixture till 25-50°F rich of peak EGT. There is normally a small reference needle on the EGT gauge which can be manually set to the peak temperature for monitoring of changes. Any change in altitude or throttle position will require a readjusting of the mixture setting.

Throttle Quadrant

The throttle, pitch, mixture and cowl flaps are sometimes referred to as the 'throttle quadrant'. This term illustrates the importance of treating these four items as one set of controls. A change should not be made to one, without considering the others. This will ensure the transition to the more complex controls of a fuel injected CSU type is made without ommissions such as leaving the cowls closed on climb out or after landing, or climbing with cruise mixture or pitch selections.

When increasing and decreasing power a flow from cowls to throttle, and throttle to cowls respectively should be followed:

↣ Increasing:
- Open Cowls;
- Enrich Mixture; (if required)
- Increase rpm;
- Increase throttle.

↣ Decreasing:
- Decrease throttle;
- Decrease rpm;
- Lean Mixture; (if required)
- Close Cowls.

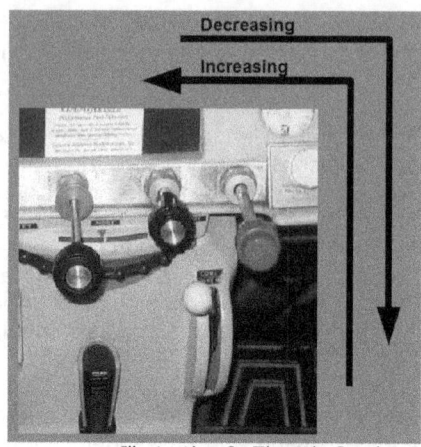

Illustration 6e Throttle Quadrant

Engine Gauges

Engine operation is controlled and monitored by the following instruments:

↣ Engine control gauges:
- Manifold pressure gauge;
- Tachometer;
- Fuel flow indicator;

↣ Engine monitoring gauges:
- Oil temperature and pressure;
- Exhaust Gas Temperature;
- Cylinder Head Temperature.

Manifold Pressure Gauge

The engine power output is most closely measured by the pressure in the inlet manifold. This pressure, the pressure of the air charge entering the cylinder, is proportional to the pressure being developed in the cylinder. In aircraft with constant speed propellers, the power is displayed by the manifold pressure gauge.

Manifold pressure is typically indicated in inches of mercury (Hg). Maximum power in a normally aspirated engine will not exceed ambient pressure.
Generally there are some losses in the engine and so the indicated manifold pressure at full throttle application is 1-2 inches below ambient pressure.
Ambient pressure can be noted by the reading on the manifold pressure gauge before start up. If the indication is more than 2 inches below ambient on application of full throttle then the engine is not developing full power and takeoff should not be attempted.
The manifold pressure has only a green arc for normal operations, there is no maximum limit, as it is not possible to exceed ambient pressure with a

normally aspirated engine. At sea level the manifold pressure at full throttle should be around 27-30".

At high elevation airfields the expected minimum take-off power can be calculated approximately by the following formula:

SEA LEVEL PRESSURE – ELEVATION/1000'-1.5" ≅ MIN TAKE-OFF POWER

For example, for take-off at airfield with elevation 4000' the expected minimum take-off power is calculated to be:

29.92" - 4" - 1.5 ≅ 24.42"

Note: turbo boosted engines will have a yellow arc and a red line on the manifold pressure, the red line is the maximum power that must be obtained by the engine (although need not be used for a normal takeoff) and must never be exceeded, the yellow arc is provided for take-off and normally has a 5 minute limitation.

Fuel Flow Gauge

Illustration 6f Manifold Pressure and Fuel Flow Gauge

The fuel flow gauge normally displays the fuel pressure converted to relative units of fuel flow. The fuel pressure is measured from a metered fuel line sourced at the fuel manifold valve.

The fuel flow gauge on the C206 is positioned opposite the manifold pressure in the same instrument case. The pressure measured is calibrated in pounds per hour, indicating approximate pounds per hour of fuel delivered to the engine. Maximum and minimum fuel pressure is also displayed in psi.

Some gauges may have different units, for example gal/hr, ensure you check the configuration before takeoff, incorrect settings will cause major power reductions on takeoff.

Tachometer

The engine-driven mechanical tachometer is located near the upper centre portion of the instrument panel. The instrument is calibrated in increments of 100rpm and indicates engine and propeller speed (direct drive) in revolutions per minute (RPM). An hour meter inside the tachometer dial records elapsed engine time and runs at full speed only when the engine develops

Illustration 6g Tachometer

full power. Hence total flight time from chock to chock is usually higher than tacho time.

IO520

IO550

Illustration 6h Example Engine Gauges

Pressure and Temperature Gauges

The oil pressure and temperature gauges are located on the left bottom side of the instrument panel. The normal operating range on both gauges is marked by a green arc.

The temperature gauge is an electric resistance type device powered by the electrical system. The pressure gauge is a mechanical direct reading device based on a "bordon tube" design.

Indications will vary from engine to engine, however (excluding the circumstances below) any deviation from the green range requires immediate action. This may include reduction in power, increasing airspeed, enriching the mixture as applicable and contemplation of a landing when practical or when required.

Temperature significantly affects pressure indications, and it is vital that temperature and pressure are considered together. In certain engines or environments, a slightly high oil pressure indication (sometimes above the green range) may be necessary when the oil is cold to prevent oil pressure becoming too low once climb or cruise operating temperatures are reached. A change of either temperature or pressure without a matching change in the other quantity may be indicative of a failure in the gauge, however fault finding should be initiated to confirm this.

Illustration 6i Temperature and Pressure Gauges (standard fitting)

CHT Gauge

The Cylinder Head Temperature (CHT) indicator, if installed, is a more accurate means of measuring the engine operating condition. It is a direct indication of engine temperature compared with oil temperature which is surrounding the engine and has inertia and damping effects. As this is one of the hottest parts of the engine, probes are often prone to failure, and may fail in a high or low position. Indications should be used in conjunction with the Oil Temperature and Pressure readings.

EGT Indicator

The Exhaust Gas Temperature (EGT) indicator is located near the tachometer. A thermocouple probe in the muffler tailpipe measures exhaust gas temperature and transmits it to the indicator. Exhaust gas temperature varies with fuel-to-air ratio, power, and RPM. The indicator is equipped with a manually positioned reference pointer.

Turbocharged Engines

Turbocharged engines use exhaust air to turn a small turbine, directly connected to a small compressor fed by ram air. The compressor, through the rotation provided by the turbine, increases the pressure of the air entering the manifold.

Engine power is directly proportional to the manifold pressure, which can now be increased significantly above that of ambient conditions, providing more power through a higher range of temperatures and pressures than the equivalent size normally aspirated engine.

The maximum takeoff manifold pressure setting for a C206 turbo engine is typically 37" Hg, which can be achieved at less than the full available power up to around 15,000ft density altitude. The normal operating range is around 15-31" the upper limit varying a little with engine model. A typical cruise power setting is around 27 inches at 2400 rpm, this can be higher or lower depending on whether speed or economy is preferred, according to the cruise performance tables for your aircraft engine installation from the POH or POH Supplement.

The increase in pressure of air results in the undesirable effect of increasing the temperature of the air, and thus increasing the engine operating temperatures. The higher temperatures and pressures and additional system demands require more complex operating procedures (see Engine Handling Tips in the Normal Operations section), and result in higher maintenance costs while providing only small gains in climb and cruise performance. For this reason turbo engines are mainly used for pressurised aircraft, or in operations out of high density altitude airfields where performance in hot and high conditions is important.

The turbine is directly controlled in proportion to the air required as dictated by the throttle setting. The only additional control in the turbo-system is the waste-gate, which opens automatically to bypass some of the exhaust air when excessive pressure is sensed in the intake manifold. It should be noted that Cessna waste-gates are typically not very reliable, and when the waste-gate is not operating as designed it is very easy to inadvertently apply power above the manifold pressure red-line ('over-boosting'). Care must be taken whenever application of full power is required, always increase the throttle slowly, holding the brakes where possible. A small or transitional (a few seconds or 1-2 inches) exceedance is permitted by the Cessna maintenance manuals to allow for the nature of the waste gate, however any large or sustained overboosting must be reported to the aircraft's maintenance organisation immediately.

A schematic of the turbo system is shown on the following page.

Turbo System Schematic

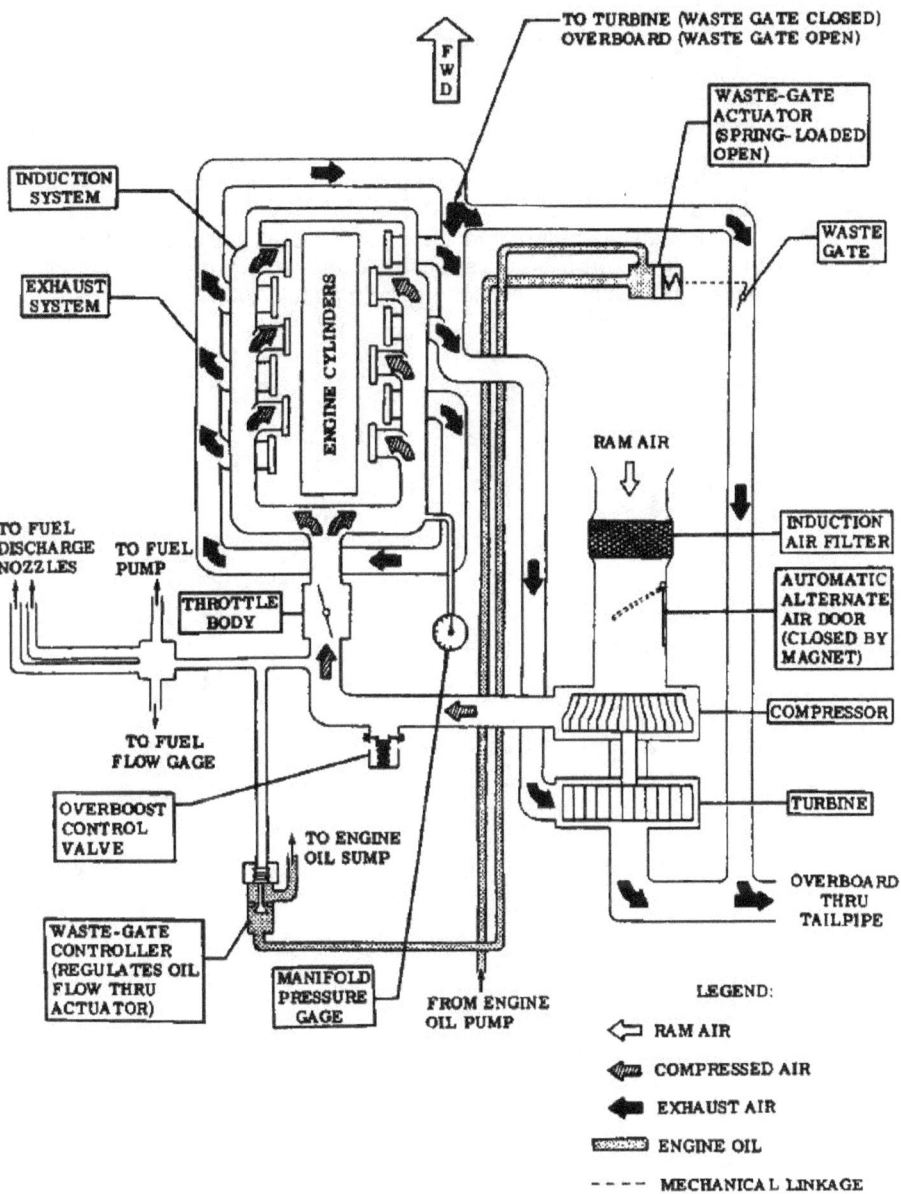

Figure 6j Turbo System Schematic

Induction System

The engine receives air through an intake on the right front of the engine cowling. An air filter, aft of the engine cylinders, removes dust and other foreign matter from the induction air. Airflow passing through the filter enters an airbox, which has a spring-loaded alternate air door.
If the air induction filter should become blocked, suction created by the engine will open the door and draw unfiltered air from inside the upper cowl area. An open alternate air door will result in an approximate 10% power lost at full throttle.
After passing through the airbox, induction air is routed either directly or via the turbocharger (for turbo engines) to the fuel/air control unit, and then ducted to the engine cylinders through intake manifold tubes.

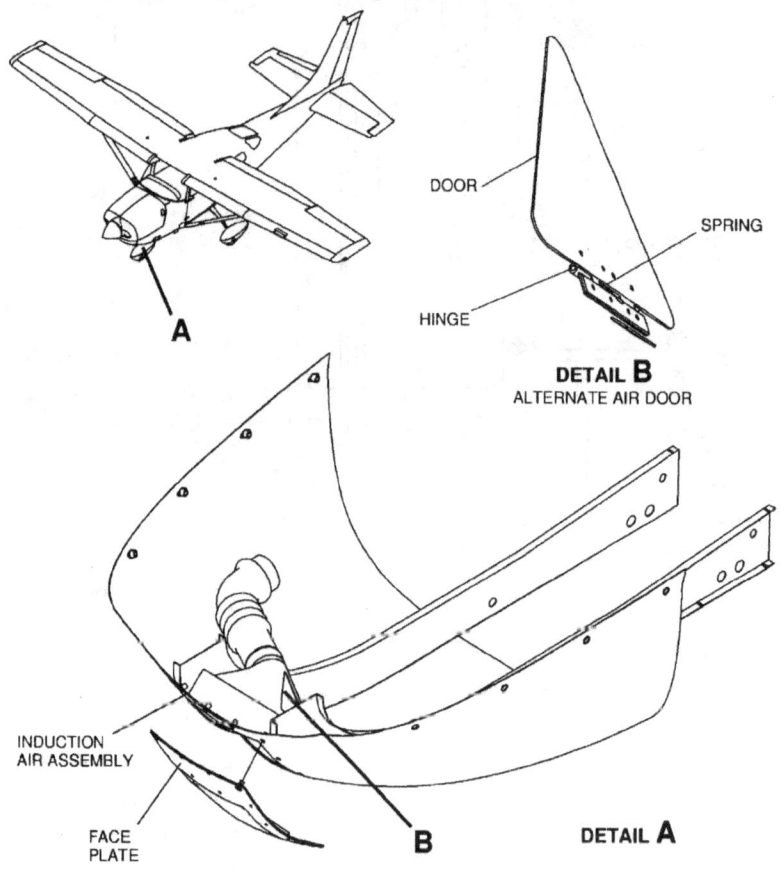

Illustration 7a Air Intake C206H

Oil System

A wet sump, pressure lubricated oil system is employed, where oil is supplied from and returned to a sump on the bottom of the engine. The capacity of the sump is 10 imperial quarts of which 2 quarts are unusable.

Oil is drawn from the sump through the engine-driven oil pump to a thermostatically controlled bypass valve. If the oil is cold, the bypass valve allows the oil to bypass the oil cooler and flow directly to the oil filter. If the oil is hot, the oil is routed to the engine oil cooler mounted on the left forward side of the engine and then to the filter. The filtered oil then enters a pressure relief valve which regulates engine oil pressure by allowing excessive oil to return immediately to the sump, while the balance of the pressure oil is circulated to the various engine parts for engine lubrication and cooling. After passing through the engine, oil is returned by gravity to the engine sump.

Because oil viscosity changes with temperature, inherent in the design of the system there will be a small change in the pressure with changes in operating temperatures, the warmer the temperature the lower the pressure. It should be noted that any large increases in temperature or decreases in pressure, or deviation from normal operating (green) range are an indication of possible malfunction. Discontinuation of the flight or landing at the nearest suitable location should be contemplated.

The oil dipstick is accessible through the inspection panel on the left side of the engine cowling. Oil should be added if the level is below 7 quarts. To minimize loss of oil through the breather, fill to 8 quarts for normal flights of less than three hours. For extended flight, fill to a maximum of 10 quarts.

Illustration 8a Oil Dipstick C206G

On older models the oil tank filler cap is separate from the oil dipstick. It is therefore required to **separately check the filler cap is secure during the preflight inspection when oil is not required.**

Access to the filler cap on older models, is through the inspection panel on the top left side of the engine cowling. The filler cap can also be seen through the left intake opening behind the propeller.

Always ensure the oil inspection panel is secure after opening.

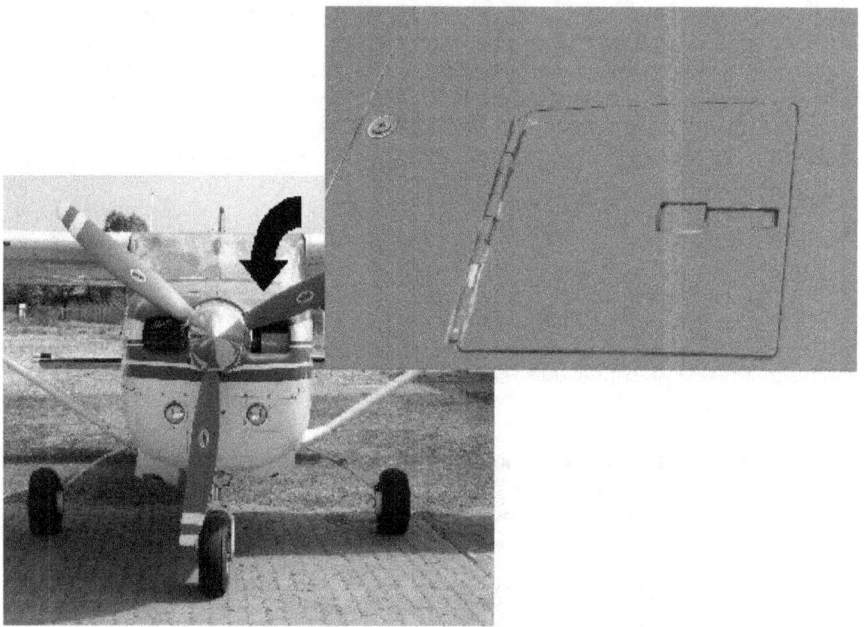

Illustration 8b Oil Filler Cap Location C206G

On the C206H the oil filler cap and dipstick are located in the same position.

Oil temperature and pressure gauges are normally fitted on the upper right part of the instrument panel (for illustrations see the Engine Section). If normal oil pressure is not indicated within 30 seconds of starting, the engine should be shut down immediately. Actions in the event of unacceptable oil pressure or temperature readings is covered in the Non Normal Flight Operations section.

Ignition System

The necessary high-tension electrical current for the spark plugs comes from self-contained spark generation and distribution units called the magnetos. The magneto consists of a magnet that is rotated near a conductor which has a coil of wire around it. The rotation of the magnet induces an electrical current to flow in the coil. The voltage is fed to each spark plug at the appropriate time, causing a spark to jump between the two electrodes. This spark ignites the fuel/air mixture.

While the engine is running, the magneto is a completely self-sufficient source of electrical energy. The aircraft is equipped with a dual ignition system (two engine-driven magnetos, each controlling one of the two spark plugs in each cylinder). A dual ignition system is safer, providing backup in event of failure of one ignition system, and results in more even and efficient fuel combustion. The left magneto is fitted on the left hand side of the engine, as viewed from the pilot's seat, and fires the plugs fitted into the top side of the left cylinders and the bottom side of the right of the cylinders, the right magneto is on the right hand side and fires the opposite plugs. The magneto selector switch is normally fitted in reverse, see below). The dual system has an added bonus of being able to isolate left and right parts for easy plug and magneto fault finding during engine run up.

Illustration 9a Magneto

Ignition and starter operation is controlled by a rotary type switch located on the left bottom side of the instrument panel. The switch is labelled clockwise: OFF, R, L, BOTH and START. When the ignition switch is placed on L (left) the left ignition circuit is working and the right ignition circuit is off and vice versa. The engine should be operated on both magnetos (BOTH position) in all situations apart from magneto checks and in an emergency. When the switch is rotated to the spring-loaded START position (with master switch in the ON position), the starter is energized and the starter will crank the engine. When the switch is released, it will automatically return to the BOTH position.

Dead Cut and Live Mag Check

It is important to realise if the ignition is live, the engine may be started by moving the propeller, even though the master switch is OFF. The magneto is self energising and does not require an external source of electrical energy.

Placing the ignition switch to OFF position grounds the primary winding of the magneto system so that it no longer supplies electrical power. With a loose or broken wire, or some other fault, switching the ignition to OFF may not ground both magnetos.
To prevent this situation, just before shutting an engine down, a "dead-cut" of the ignition system should be made.

The dead-cut check is made by switching the ignition momentarily to OFF and a sudden loss of power should be apparent. This is carried out most effectively from R, not from Both, to prevent inadvertent sticking in OFF.

On start up, a live mag check is performed, to ensure both magnetos are working before taxi. This is not a system function check detailed below, as the engine is still cold and plugs may be fouled, rather just a check to ensure both magnetos are working by switching from Both to L, then R, and back to Both, noting a small drop from Both in L and R positions. A dead-cut check may be carried out at the same time.

The engine will run on just one magneto, but the burning is less efficient, not as smooth as on two, and there is a slight drop in RPM.

The magneto check to confirm both magnetos and plugs are operational should be made at 1700rpm as follows:
↱   Move ignition switch to R position and note the RPM;
↱   Then move switch back to BOTH to clear the other set of plugs;
↱   Move switch to the L position, note the RPM and return to BOTH position.

RPM drop should not exceed 150rpm on either magneto or show greater than 50rpm difference between magnetos.

An absence of RPM drop may be an indication of faulty grounding of one side of the ignition system, a disconnected ground lead at the magneto, or possibly the magneto timing is set too far in advance. Excessive drop or differential normally indicates a faulty magneto.

Fouled spark plugs (lead deposits on the spark plug preventing ignition) are indicated by rough running usually combined with a large drop in RPM (i.e. one or more cylinders not firing). This is due to one of the two plugs becoming fouled, normally the lower plug. Plug fouling, if it is not excessive, can normally be burnt off.

Run the engine at a correct or slightly lean mixture setting and a high power setting (+/-2000rpm) for a few minutes, caution engine temperatures and surrounds. If the mis-firing still exists then the spark plug may be faulty (or the carbon build up is too excessive to remove) and the spark plug needs to be replaced.

Cooling System

The engine cooling system is designed to keep the engine temperature within those limits designed by the manufacturer. Engine temperatures are kept within acceptable limits by:

- The oil that circulates within the engine;
- The air cooling system that directs and circulates fresh air around the engine compartment;
- Cowl flaps that increase or decrease the flow of air through the engine compartment;
- An air cooled oil cooler/heat exchanger mounted at the front of the engine.

The engine is air-cooled by exposing the cylinders with their cooling fins, and the oil cooler into the main airflow. Air for engine cooling enters through two openings in the front of the engine cowling. The cooling air is directed around the cylinders and other areas of the engine by baffling, and is then exhausted through an opening at the bottom aft edge of the cowling.

Air cooling is least effective at high power and low airspeed, for instance on take-off and climb. At high airspeed and low power, for instance on descent, the cooling might be too effective. It is therefore important to monitor the cylinder-head temperature gauge throughout the flight, and also on the ground when air-cooling will be poor.

Oil Cooler

The oil cooler greatly assists in reducing the high operating temperatures of the high performance engine. Both the Bonaire and the turbo engines run at slightly higher operational speeds and a larger oil cooler can be fitted to assist with engine life and operating efficiency. The turbo model will normally run at significantly higher temperatures than the same engine operating without a turbo due to the higher manifold pressures and heat generated by the turbocharger. High temperatures alone should be cause for concern, provided the oil temperature remains below the maximum, and providing there is no sudden or significant change from the normal operating temperature range for the ambient conditions and power setting.

Illustration 10a Oil Cooler

Operation of Cowl Flaps

The cowl flaps should be thought of as part of the power quadrant. Whenever a change in power is selected it should be made from right to left for increasing, beginning with the cowl flaps and finishing with throttle application, or from left to right for decreasing power, finishing with the cowl flaps only after setting the throttle, pitch and mixture. Cowls may be selected to open, closed or partial settings.

Cowl flaps should be open whenever the temperatures require for assisted cooling. This includes:
- high power operations (takeoff and climb);
- whenever the air cooling is insufficient (ground operations);
- high ambient temperatures or with high indicated engine temperatures.

Cowl flaps should be normally selected closed for:
- cruise;
- descent and approach;
- low power operations.

During cruise, after extended climbs or in high outside temperatures it may be required to select cowls to half or to leave the cowls open until the engine temperature has stabilised.

Cowls may be left open for circuit operations, however it is found that more stable temperatures are achieved when cowls are selected to half or closed after reaching circuit altitude. This also reinforces a routine to remember to open and close the cowls whenever changing power and to open cowls when completing checks before landing.

Cowls should be opened on short final, to ensure cooling is available for ground operations or in preparation for a go-round. This should not be carried out on base for fear of forgetting as too much cooling will result while the engine is still at low power settings with cooling airflow. This check will be part of your final approach checks, and is rechecked on the ground in the after landing checks, or on climb out for the missed approach in the after take off checks.

Other Cooling Methods

If excessive temperatures are noted in flight, additional cooling of the engine can be provided by:
- Enriching the mixture (extra fuel has a cooling effect in the cylinders);
- Reducing the engine power;
- Increasing the airspeed;

The propeller spinner in addition to streamlining and balance is a director for the cooling air, and so the aeroplane should generally not be operated without the spinner.

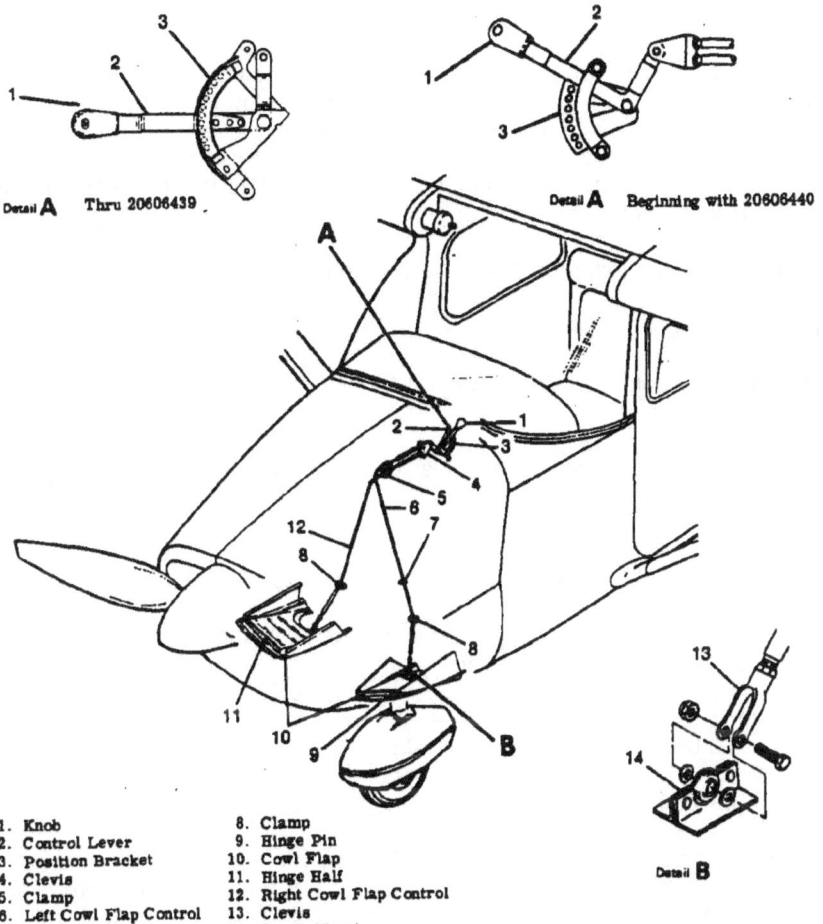

1. Knob
2. Control Lever
3. Position Bracket
4. Clevis
5. Clamp
6. Left Cowl Flap Control
7. Clamp (Left Side Only)
8. Clamp
9. Hinge Pin
10. Cowl Flap
11. Hinge Half
12. Right Cowl Flap Control
13. Clevis
14. Shock Mount

Illustration 11a Cowl Flap Diagram

Fuel System

The fuel system consists of two vented fuel tanks (one in each wing), two fuel reservoir tanks, a fuel selector valve, auxiliary fuel pump, fuel strainer, engine-driven fuel pump, fuel/air control unit, fuel manifold, and fuel injection nozzles.

Fuel flows by gravity from tanks to two reservoir tanks, to a three-position selector valve, through a bypass in the auxiliary fuel pump (when it is not in operation), and through a strainer to an engine-driven fuel pump. The engine-driven pump delivers the fuel to the fuel/air control unit where it is metered and directed to a manifold which distributes the metered fuel/air mixture to each cylinder.

A schematic of the fuel system can be seen on the following page.

Fuel Tanks

The main fuel tanks are either integral tanks (late models) or bladder tanks (early models).

Models may differ slightly in the fuel capacity, typical fuel tank capacities are:

- Standard Tanks 1964-1977:
    61USG, 29.5 USG usable each side (65USG Total);
- Long Range Tanks 1964-1977:
    80 USG, 38 USG usable per side (84USG Total);
- Standard Tanks 1977-1978:
    59USG, 29.5 USG usable each side (63USG Total);
- Long Range Tanks 1977-1978:
    76 USG, 38 USG usable per side (80USG Total);
- Wet Wing (Integral) Tanks (1979 on):
    88 USG, 44USG usable per side (92USG Total).

It is important to remember with regard to fuel planning, that the amount of fuel we can put into fuel tanks is limited by the volume of the tanks, and therefore usable fuel is always provided in volume, such as gallons and litres. However, the engine is only sensitive to the mass of fuel, and not to the volume. The engine will consume a certain mass (lbs or kgs) of fuel per hour.

Fuel System Schematic

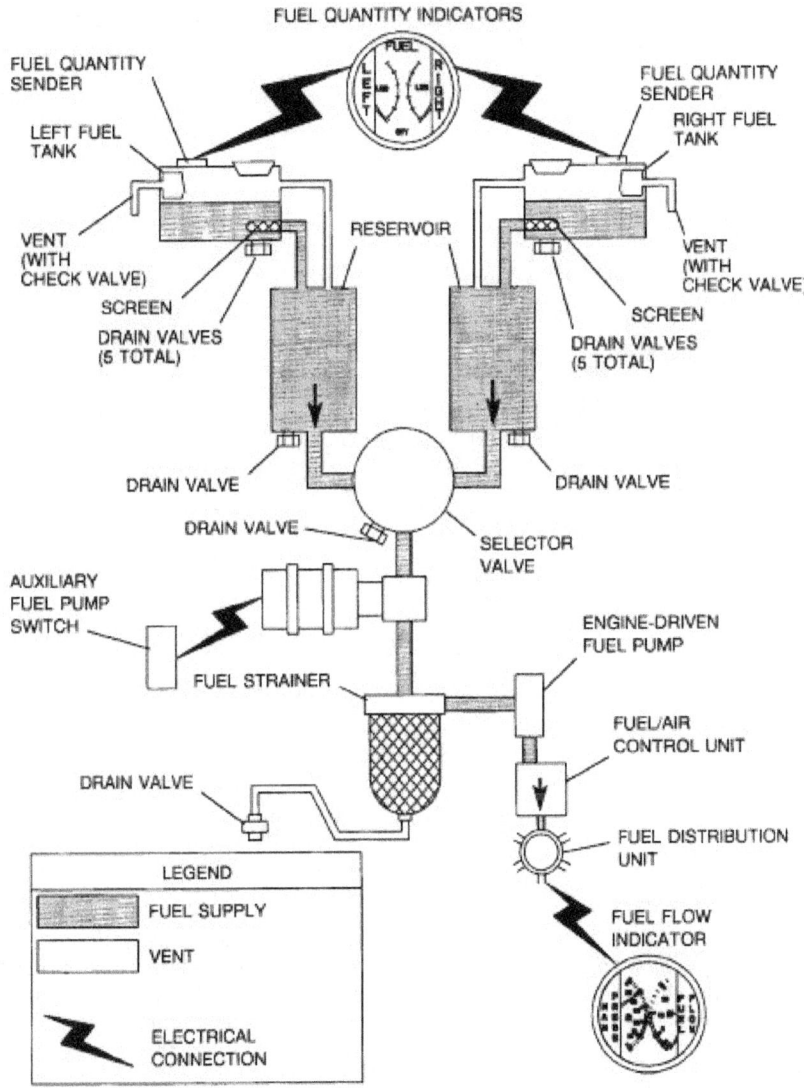

Illustration 12a Fuel System Schematic C206H

Fuel has a wide variation in specific gravity (weight of fuel per volume) mostly depending on temperature and type of fuel. Therefore, variations in specific gravity of fuel can have a significant effect on the mass of fuel in the tanks and therefore the range and endurance. For practical purposes the specific gravity of Avgas is taken as 0.72 kgs/lt.

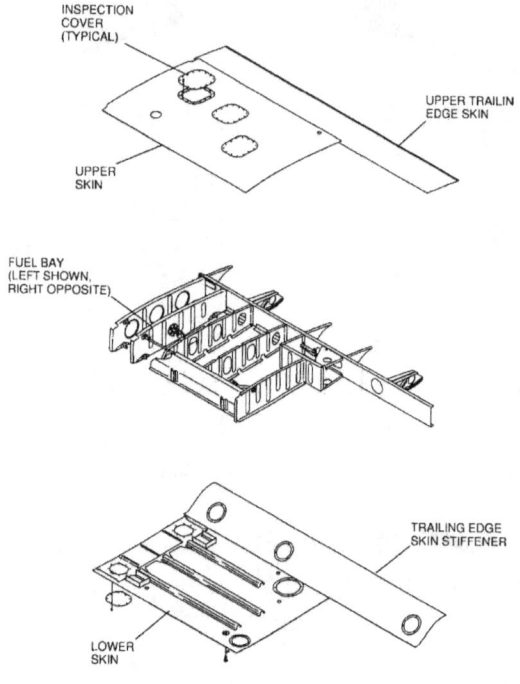

Illustration 12b Integral Fuel tank

Bladder Tanks

The majority of C206's produced (up to 1979) before production ceased, were fitted with bladder tanks. That is the tank consists of a rubber bladder fitted inside the wing, instead of the preferred integral sealed wing tanks. Bladder tanks have a tendency to develop wrinkles and trap water in folds. If this occurs even though the tanks have been drained there may be still water present. Shaking the wings to ensure water is dislodged from creases and draining again after settling is recommended, especially if the aeroplane has been standing outside overnight in moist or wet weather.

Some models are additionally susceptible to moisture intake due to their retaining the original flush style fuel caps. These caps have a small indent where water may collect and seep through the vent. Although these caps were fitted by Cessna, most have been replaced by recommended alternatives.

Tip Tanks

Optional tip tanks may be installed for extended range. Tip tanks do not feed the engine directly, the fuel has to be transferred into the main tank by a transfer pump. Each tip tank has a transfer pump and a fuel quantity gauge, which are normally mounted on the left side of the control column.

Illustration 12e Fuel Tip Tank Gauges

The tip tank installation will have specific operating instructions noted in the flight manual or flight manual supplement.

Points to remember on tip tank operation are:
- The main tanks must be 15 to 20 Gallons below full before transfer is commenced, to ensure sufficient space for the fuel being transferred, or fuel will be vented overboard;
- Transfer may be completed only when the tank is not in use (except in an emergency which requires deviation from normal operating procedures);
- If the tip tank installation provides for a higher maximum weight limit (3800lbs), the higher weight is normally only applicable if the tanks are more than half full. Speeds and climb performance for the additional weight are the same, however distances are increased by 10%;
- Although no additional take-off and landing speeds are available, the additional wing length creates additional lift, and so when flown at the same speeds increases the tendency for the aircraft to float during the flare.

Fuel Selector and Shut-off Valve

The fuel valve is located on the floor of the cockpit between the pilot and co-pilot seats.

The typical installation has three positions: LEFT, RIGHT and OFF. Fuel cannot be used from both fuel tanks simultaneously. A methodical procedure for fuel balancing must be adopted to ensure fuel is used evenly from both tanks, (see more under Cruise, in the Normal Operating Procedures section).

The C206H has LEFT, RIGHT, BOTH and OFF positions on the fuel selector. BOTH must be used for takeoff, and should be used for all normal operations, except when fuel balancing is required. It is important if operating in the both position to monitor fuel usage to ensure the remaining quantity in each tank remains balanced, and select Left or Right if there is any significant variation.

It is also very important to remember to position the fuel selector to LOW wing or to OFF when parked. Never leave the aircraft parked in the BOTH position, as fuel will flow from the high tank to the low tank through the BOTH position selector and out of the wing overflow. This process can cause significant fuel loss when parked on uneven ground, even where the gradient is barely noticeable.

Refuelling

Fuel starvation or contamination, in both the Cessna integral and bladder tank models, has often resulted from failure to correctly fill and check the fuel tanks.

🍀 The following precautions must be noted.

Integral tanks: With integral tanks it is possible to have airlocks in the interlinked compartments, resulting in the wing tank appearing full but containing less than full capacity. To avoid problems during filling, ensure the aircraft is level, and refuelling is carried out slowly, allowing the fuel time to settle. Where in doubt, rock the wings and allow a few minutes to settle before filling the last few gallons. A fuel monitoring program requiring recorded checks in the flight log, before and after filling, will also aid in eliminating problems.

Bladder Tanks: It is common for water to be trapped in crevices of the bladders. To move any trapped water to the drain points also requires gentle shaking or rocking of the wings. Thereafter allow the fuel and water to resettle and recheck the fuel drains. The fuel should be allowed to settle approximately 15 minutes, longer where possible, before draining. Ideally fill up the previous night.

Filler Cap Quantity

Late models provided, with 88USG tanks contain a filler neck which provides a convenient method of filling the tanks to approximately ¾ capacity, or 30USG useable per side. Where a filler neck is provided, a quantity placard is usually required by the POH, near the fuel tank cap, indicating the full and filler neck quantity.

Fuel Venting

Fuel system venting is essential to system operation and is necessary to allow normal fuel flow or pressure venting as fuel is used. Blockage of the venting system will result in a decreasing fuel flow and eventual engine failure through fuel starvation.

Venting is accomplished by crossover vent lines, one from each tank, in each forward door post, and by wing tip vents which are equipped with check valves. The fuel filler caps are also equipped with vacuum operated vents which open, allowing air into the tanks, should the tank vent lines become blocked.

The vent line opens to the highest part of the tank. Although there is a non return check valve to prevent loss of fuel through the vent, to allow for expansion of fuel through heating there is a small hole provided in the check valve. It is normal if the tanks are full to see a small amount of overflow fuel leaking through the fuel vent.

Fuel Drains

The fuel system is equipped with drain valves to provide a means for examination of fuel in the system for contamination and grade.
Fuel should be examined before the first flight of every day and after each refuelling, by using a sampler cup to drain fuel from the wing tanks, sump, and by utilising the fuel strainer drain under an access panel on the left side of the engine cowling. Quick-drain valves are also provided for the fuel reservoir tanks to provide a simple means of draining the tanks.

Fuel drains are spring-loaded valves at the bottom of each fuel tank. There is usually a drop in air temperature overnight and, if the tank is not full, the fuel tanks' walls will become cold and there will be a lot more condensation than if the tanks were full of fuel. The water, as it is heavier than fuel, will accumulate at the bottom of the fuel tanks.

If water is found in the tank, fuel should be drained until all the water has been removed, and wings should be rocked to allow any other water to gravitate to the fuel strainer drain valve (see note regarding bladder tanks under the Refuelling section above).

There are normally two under wing drains, however some models may have additional drain points installed often to attempt to combat the moisture problems with bladder tanks.

Illustration 12c Fuel Reservoir Drains

There are normally also two drains underneath the fuselage directly beneath

the forward cabin. On models with a pod, they are fitted inside the pod, and particular care should be taken with baggage in the case of spilled fuel or leaks.
Colloquially known as belly or sump drains, these drains are correctly termed reservoir drains as each drains from a small reservoir close to the fuel selector. These reservoirs are the lowest point in the fuel system, and therefore prone to water content, especially if the aircraft is standing for long periods.

Fuel drains are sealed by rubber 'O' rings. These rings need periodic replacement, evidence is sometimes indicated by black rubber particles in the fuel sample or fuel staining (green or blue dye from the fuel) around the strain point, development of a slow leak or improper seating of the valve after checking the fuel (i.e the valve does not close properly). In any of these cases the situation should be reported to the maintenance provider.

A fuel strainer is normally mounted on the left forward side of firewall in the lower engine compartment, with the quick-drain valve adjacent to the oil dipstick and is accessible through the oil dipstick door. Care should always be taken to ensure the quick drain valve is selected to fully closed after straining.

All drains should be checked for signs of leakage (fuel dye stains or dripping), and deterioration of rubber seals or bladders (small black pieces in fuel sample).

Fuel Measuring and Indication

Fuel quantity is measured by two float-type variable resistance fuel quantity transmitters (one in each tank), and indicated by two electrically-operated fuel quantity indicators on the lower left portion of the instrument panel.

The full position of float produces a minimum resistance through the transmitter, permitting maximum current flow through the fuel quantity indicator and maximum pointer deflection. As the fuel level is lowered, resistance in the transmitter is increased, producing a decreased current flow through the fuel quantity indicator and a smaller pointer deflection. An empty tank is indicated by a red line and letter E. When an indicator shows an empty tank, approximately 0.5 gallons remain in the tank as unusable fuel.
The float gauge will indicate variations with changes in the position of fuel in the tanks and cannot be relied upon for accurate reading during skids, slips, or unusual attitudes.
Considering the nature of the system, take-off is not recommended with less than 1 hour total fuel remaining. Fuel quantity should always be confirmed by dipstick during the preflight inspection and on intermediate stops enroute.

To permit better payloads it is often required to reduce the fuel loading. A quick method of reduced fuelling can be accomplished by filling each tank to the bottom edge of the fuel filler neck (the aluminium cylinder protruding from the inside of the filler cap). This results in a reduced fuel load of approximately 195 pounds in each tank. This provides approximately 4 hours endurance, or 3 hours safe flight time. If less than 3 hours flight time is planned fuel can be uploaded using a suitably calibrated dipstick.

Auxiliary Fuel Pump and Priming System

The auxiliary fuel pump is located forward of the fuel reservoir(s). It is connected in line with engine driven pump, therefore, all fuel must flow through auxiliary pump internal by-pass valve.

The auxiliary fuel pump switch is a yellow and red split-rocket type switch located on the left lower pilot instrument panel.

The auxiliary fuel pump is used for priming, vapor purging and engine driven pump failures.

The yellow right half of the switch is labelled "START" and "ON" and is used for normal starting and minor vapour purging during taxi. The red left half of the switch is labelled "EMERG" and "HI" is used in the event of an engine-driven fuel pump failure during take-off or high power operation. The "HI" position can also be used, when required, for extreme vapour purging (if the ON selection is insufficient), purging the fuel lines prior to starting and when the low side is insufficient for priming.

With the yellow switch in ON position, the pump operates at one of two flow rates that are dependent upon the setting of the throttle. With throttle open to a setting, the pump operates at a high capacity to supply sufficient fuel flow to maintain engine demand in flight. When the throttle is moved toward the closed position, the fuel pump flow rate is automatically reduced to prevent an excessively rich mixture. Maximum fuel flow is produced when the left half of the switch is held in the spring-loaded "HI" position. In the "HI" position, an interlock with the switch automatically trips the remaining (yellow) half of the switch to the ON position. When left (red) half of the switch is released, the right (yellow) half will remain in the ON position until manually returned to the OFF position.

If the auxiliary fuel pump switch is placed in the ON position with the master switch ON, when the engine is not running and the mixture is in any position except idle cut off, the intake manifold will be flooded. Care should be taken not to inadvertently activate the fuel pump when not required.

Unlike a low wing fuel injected engine, the auxiliary fuel pump is not required for redundancy purposes during take-off or low level operations, due to the assistance provided by the pressure from the high wing gravity feed fuel supply.

Illustration 12d Auxiliary Fuel Pump Bonaire

Bonaire placards a warning that the fuel HI side of the fuel pump should only be used for priming and EDP failures only, some early manuals indicate using the fuel pump on high for severe vapour surges and or for vapour locks. The important point to remember is that after start the HI selection should not be used for any normal operations, hence the switch is sprung loaded to OFF. Pilots transitioning from a low wing should also remember that the ON side of the pump is NOT required in normal operations.

Priming is achieved by use of the auxiliary fuel pump either the ON or HI position depending on the priming requirements. In both positions, the throttle can then be used to control how much fuel is given to the engine depending on the engine model, the ambient and engine temperatures.

Priming the engine is normally required when starting a cold engine, when the fuel in the line is reluctant to vaporize. Priming should be carried out immediately prior to starting. If priming is carried out too early the fuel is ineffective in the start cycle, but effective in washing oil from the cylinder walls and causing additional frictional wear on start.

With the mixture in the idle cut off position, the auxiliary pump may be used to clear vapour locks and ensure there is adequate fuel in the fuel lines for starting. This process is called "purging" or "priming the fuel lines". The most effective means to prime the lines is with the throttle fully open and the pump selected to high (both switches) for approximately 10 seconds. Always check the mixture control is fully closed (idle cut off position), and no fuel flow is evident, to ensure no fuel is reaching the engine.

Priming on Continental versus Lycoming

It is important to note that different engines can have very different priming requirements for start. The biggest difference in the C206 model series is between the Continental engines on older models and the Lycoming engines on the restarts. Whilst Continental engines need much more priming, even when hot, Lycoming engines easily flood and priming should be applied with care.

More on priming and starting, including the different techniques for each engine, can be found in the Normal Operations section.

Vapour Locks in the Fuel System

The Cessna 206 is occasionally known to experience vapour locks in the fuel system. If you are lucky this may only cause minor problems during starting or taxi (see more under starting in the flight operations section).
It is possible for a vapour lock in the fuel system to cause engine surges or fuel starvation during flight under normal operations.
This situation is most common on an extended climb, but could occur at any time, and is characterised by an engine surge/stoppage, which later appears to have no cause.

At any sign of engine surge or fault, especially if accompanied by fuel flow fluctuations, the auxiliary fuel pump should be selected on to LOW, or if surging continues, held in the spring loaded HIGH position, as described in the manufacturer's handbook, and an alternate tank selected.

※ In the majority of Cessna 206 models, the engine driven pump feeds back into the selected reservoir tank. This hot return fuel can cause vaporisation in the reservoir tank, a common reason fuel system surges and stoppages. Selection of the auxiliary fuel pump (which draws from the reservoir tank) can make the situation worse, without selecting a new tank to ensure a new cooler supply of fuel to the engine.

Fuel Injection System

The fuel injection system is a low pressure system injecting fuel into the inlet valve of each cylinder.

The fuel injection system contains:
↳ **fuel injection pump** - rotary vane type engine driven fuel pump providing pressure for fuel injection;
↳ **auxiliary pump** - electrical standby fuel pump for starting and engine driven pump failure;
↳ **fuel air control unit** - controls the air intake and meters the fue pressure to obtain the desired fuel air ratio, including:

- throttle /fuel and mixture control linkages;
- a metering unit;
- a manifold valve - central point for distribution to cylinder discharge nozzles (fuel inlet to cylinder).

A schematic of the fuel injection system is shown below.

Fuel Injection System Schematic

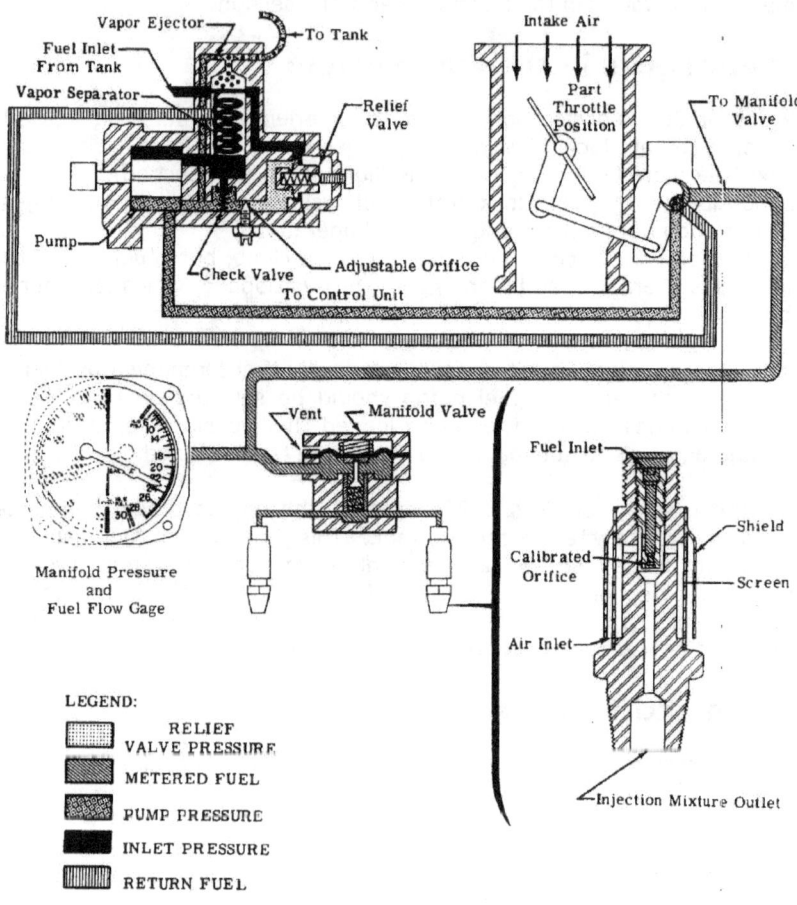

Illustration 12f Fuel Injection Schematic C206G

Electrical System

Electrical energy for the aircraft is supplied by a direct-current, single wire, negative ground, electrical system with a lead acid battery.

The system is either:

For models before 1965:
✦ 12 Volt, 33 amp-hours Battery;
✦ 14 Volt 35 or 50 Amp Generator.

For models after 1965:
✦ 12 Volt, 33 amp-hours Battery;
✦ 14 Volt 52 or 60 Amp Alternator.

Or for models after 1973:
✦ 24 Volt, 12.75 or 15.5 amp-hours Battery;
✦ 28 Volt, 60 or 95 Amp Alternator.

Aircraft equipped with the G1000 (glass cockpit) additionally have a 24V standby battery, for operation of the G1000's flight instruments, navigation and communications and engine instrument for approximately 30 minutes after failure of the main battery or on pilot selection following an alternator failure.

Battery

The battery supplies power for starting and furnishes a reserve source of power in the event of alternator or generator failure.

Battery capacity in amp-hours provides a measure of the amount of load the battery is capable of supplying. This capacity provides a certain level of current for a certain time. A 25 amp-hour battery is capable of steadily supplying a current of 1 amp for 25 hours and 5 amp for 5 hours and so on.

The battery may be located under the floor in the the rear baggage, under the floorboards beneath the pilot's seat (1961 models), or most commonly under the engine cowl, behind the firewall.

Alternator/Generator

An engine-driven alternator or generator is the normal source of power during flight and maintains a battery charge, controlled by a voltage regulator/alternator control unit.

The charging system capacity (28 or 14 volt), is the output from the alternator after voltage regulation. This is always slightly more than the battery (24 or 12 volt) to ensure continuous charge to the battery when using the electrical system in normal operations.

The generator if installed is either 35 or 50 Amps, alternators may be 14 Volts and 52 or 60Amps, or on the later models 28 Volts, 60 or 95 Amp.External Power

An external power source receptacle is offered, to supplement the battery alternator system, for starting and ground operation. An External power receptacle for connection of a secure power supply for starting is provided.

The external power receptacle is either on the left front nose cowling or on the rear cowling near the baggage compartment door depending on the model.

Electrical Equipment

On the Cessna 206, the following standard equipment requires electrical power for operation (there may be additional optional equipment which uses electrical power):
- Fuel quantity indicators;
- All internal and external lights and beacon, including warning lights;
- Pitot heat;
- Wing flaps;
- Starter motor;
- All radio and radio-navigation equipment;
- Turn coordinator and any other electrically powered instruments.

System Protection and Distribution

Electrical power for electrical equipment and electronic installations is supplied through the split bus bar. The bus bar is interconnected by a wire and attached to the circuit breakers on the lower, centre of the instrument panel.

The circuit breakers are provided to protect electrical equipment from current overload. If there is an electrical overload or short-circuit, a circuit breaker (CB) will pop out and break the circuit so that no current can flow through it.

It is normal procedure (provided there is no smell or other sign of burning or overheating) to reset a CB once only, after a cooling period, by pushing it back in.
⚠ Any further attempt to reset may cause system damage or fire.

Most of the electrical circuits in the aeroplane are protected by "push-to-reset" type circuit breakers. However, alternator output and some others are protected by a "pull-off" type circuit breaker to allow for voluntary isolation in case of a malfunction.
Electrical circuits which are not protected by circuit breakers are the battery contactor closing (external power) circuit, clock circuit, and flight hour recorder circuit.

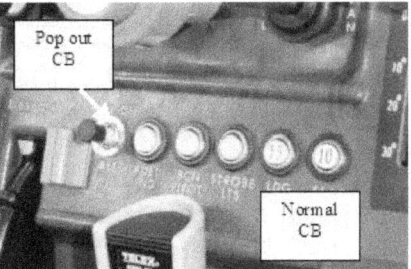

Illustration 13a Circuit Breakers

These circuits are protected by fuses mounted adjacent to the battery and are sometimes termed "hot wired or hot bus" connections because they are directly wired to the battery.

The master switch controls the operation of the battery and alternator system. The switch is an interlocking split rocker type, with the battery mode on the right hand side and the alternator alternator mode on the left hand side. This arrangement allows the battery to be on line without the alternator, however, operation of the alternator without the battery on the line is not possible.

The switch is labelled BAT and ALT and is located on the left-hand side of the instrument panel. Continued operation with the alternator switch OFF will deplete the battery power. If the battery power becomes too low the battery contactor will open, removing power from the alternator field, and prevent the alternator from restarting. This is important to remember if you are starting an aeroplane by other means because of a flat battery.

Illustration 13b Master Switch

The ammeter, located on the upper right side of the instrument panel, indicates the flow of current, in amperes, from the alternator to the battery (charge) or from the battery to the aircraft electrical system (discharge).

When the engine is operating and the master switch is ON, the ammeter indicates the charging rate applied to the battery. When the ammeter needle is deflected right of centre, the current flows into the battery and indicates the battery charge rate. When the ammeter needle is deflected left of centre, the current flows from the battery and the battery is therefore discharging.

With battery switch ON and no alternator output, the ammeter will indicate a discharge from the battery, because the battery is providing current for the electrical circuits that are switched on.

If the alternator is ON, but incapable of supplying sufficient power to the electrical circuits, the battery must make up the balance and there will be some flow of current from the battery. The ammeter will show a discharge. In this case, the load on the electrical system should be reduced by switching off unnecessary electrical equipment until the ammeter indicates a charge.

A charge rate from the alternator to the battery will be shown due to fluctuations in the supply and demand and lags in the voltage regulation control. Indication of charge from the system to the battery more than temporarily, may indicate more serious problems and should be checked out immediately. Excessive charging may damage the battery, or essential electrical equipment and can cause an electrical fire.

An alternator failure may be detected by a red warning light near the ammeter labelled, HIGH-VOLTAGE (1969-1978), on early models or LOW VOLTAGE on later models, or by the VOLTS annunciator on the C206H.

The switch was labelled HIGH VOLTAGE on early models as illumination would occur after the alternator was removed from the circuit due to an over-voltage condition, however the light will come on whenever the alternator is removed from the circuit. Although there is a slight difference in the components of the two systems, the basic function is the same, and is described more accurately by the later name.

In models with a generator, there is a light labelled GEN, which indicates when the generator is supplying insufficient power to the battery, which will illuminate steady if the generator is removed from the circuit, and will also often flicker on at idle or in situations requiring high electrical load.

Models without an over-volt sensor (1966-1969) do not have a warning light, although one may have been installed later as an optional modification. On these models, system protection is provided by means of a circuit breaker, and the only means of determining if the alternator is offline is by the ammeter discharge.

In the event an over-voltage condition occurs, the over-voltage sensor, if installed, automatically removes the alternator field current and shuts down the alternator. The red warning light will then turn on, indicating to the pilot that the battery is supplying all electrical power. This condition can be confirmed by a discharge on the ammeter.

The over-voltage sensor may be reset by turning both sides of the master switch OFF and back ON again. If the light extinguishes, the over-volt condition was transient, however if the light illuminates again, a malfunction has most likely occurred, and the remainder of the flight will be with an electrical supply from battery source alone.

If illumination of the warning light is due to a temporary under-volt, the switch does not need to be recycled, as an over-voltage condition has not occurred to de-activate the alternator, and the light will go out once the voltage returns to normal.

The warning light may be tested by momentarily turning OFF the ALT portion of the master switch.
In all aircraft, although unlikely, it may be possible to have an over-voltage situation which does not trip the protection mechanisms, for example where the protection mechanism has been hard-wired by a large surge in voltage, or where a high voltage protection mechanism is faulty.

This will only be evident by an excessive rate of charge on the ammeter. In this case the alternator must be shut off as soon as possible to prevent damage to the battery or electrical system, including a possible electrical fire. Reducing the electrical load before attempting to reset may assist in rectifying the situation, however due to the large variations in systems and the non time-critical nature of electrical faults, for secondary actions the POH must be referred to.

For more information on electrical system malfunctions refer to Non-normal Procedures section.

On the following page a schematic of the electrical system can be seen.

Electrical System Schematic

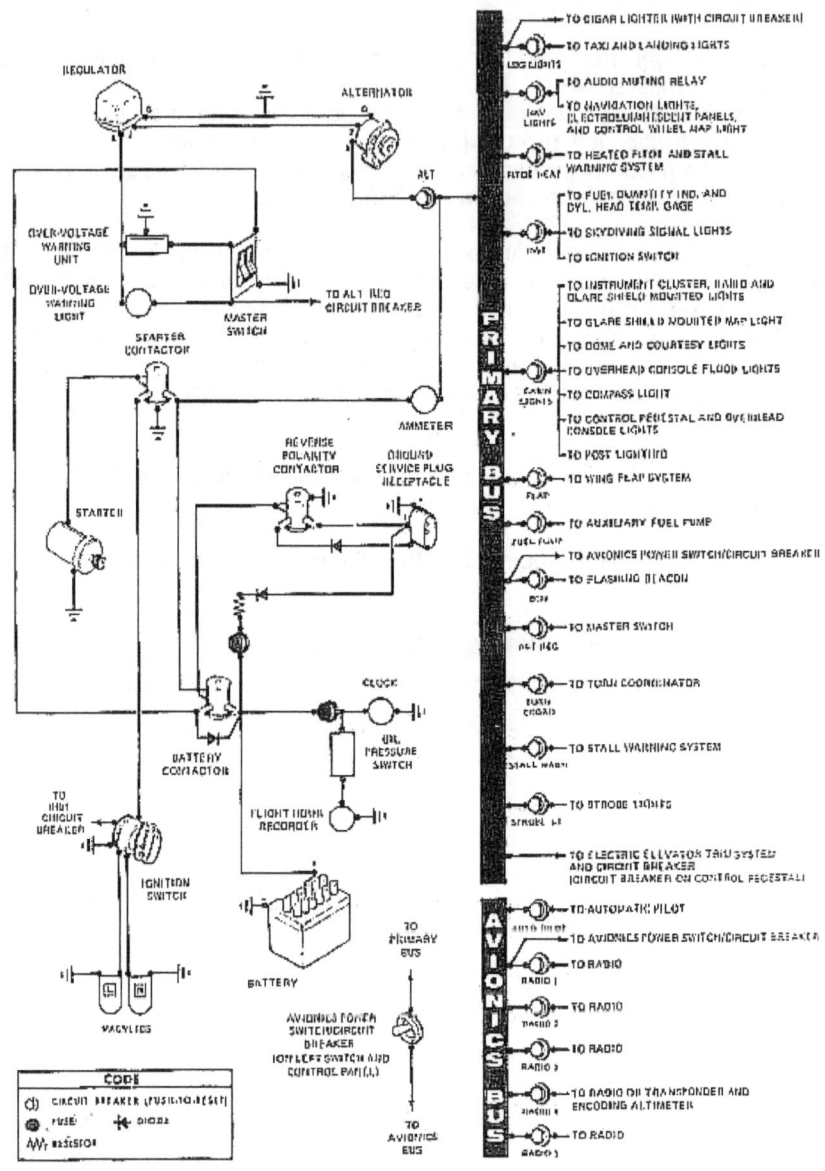

Illustration 13c Electrical Schematic

Flight Instruments and Associated Systems

The Cessna 206 with conventional instrumentation (non-glass), is normally equipped with the following standard flight instruments:

- **Attitude Indicator:** A gyro driven instrument, normally operated by the vacuum system, providing a visual indication of flight attitude. A knob at the bottom of the instrument is provided for adjustment of the miniature aeroplane to the horizon bar;
- **Directional Indicator:** A gyro driven instrument, normally operated by the vacuum system, which displays aeroplane heading on a compass card. A knob on the lower left edge of the instrument is used to adjust the compass card to the compass;
- **Airspeed Indicator:** Operated by dynamic and static pressure, and may be calibrated in knots or miles per hour. The instrument also displays airframe/aerodynamic limitation markings in the form of white, green and yellow arcs and a red line;
- **Altimeter:** Operated by static pressure, depicts aeroplane altitude in feet. A knob near the lower left edge of the instrument provides adjustment of the barometric scale to the current altimeter setting – QNH/QNE/QFE;
- **Vertical Speed Indicator:** Operated by static pressure and depicts the rate of climb or descent in feet per minute;
- **Turn and Slip Indicator:** A gyro driven instrument, normally operated by electric power, providing for rate of turn indication (instrument does not indicate bank), the slip indicator (or "ball") is operated purely by gravity in a similar way to a water level.

Illustration 14a Flight Control Instruments

The G1000 equipped aircraft have the same instrumentation, however these instruments are displayed in digital format on the Primary Flight Display (PFD) screen, and some instruments' sources are different.

G1000 Data Source Diagram

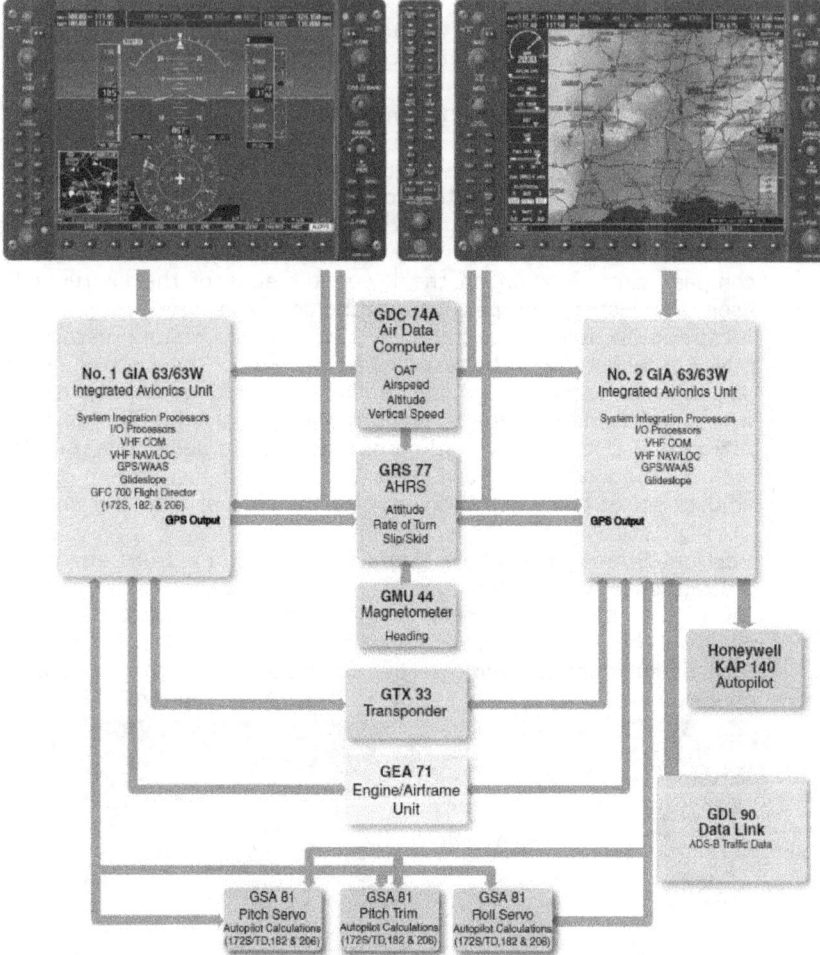

Illustration 14b Garmin 1000 Schematic

Directional and attitude information is obtained from the Attitude Heading Reference System (AHRS), and magnetic information is sourced from a magnometer. The pitot-static instruments on the PFD still receive information from the pitot and static tubes, but this information is first fed to the air-data computer (ADC) for conversion into digital format. The pitot-static system also supplies a conventional altimeter and airspeed indicator, and a suction pump operates a conventional gyro driven artificial horizon, for the standby instrumentation.

Pitot-Static Instruments

The pitot-static system supplies static and dynamic air pressure to the airspeed indicator and static air pressure to the vertical speed indicator and altimeter. The system is composed of a pitot tube mounted on the lower surface of the left wing, two external static ports, one on each side of the aft fuselage, and the associated plumbing necessary to connect the instrument to the sources.

The heated pitot system consists of a heating element in the pitot tube, and a switch labelled PITOT HT on the lower left side of the instrument panel. When the pitot heat switch is turned ON, the element in the pitot tube is heated electrically to avoid ice building on the pitot tube when operating in icing conditions.

The pitot tube and static vent must not be damaged or obstructed, as this will cause false readings from the relevant flight instruments. This can have disastrous results, especially to IFR flight.

Both the pitot and static ports should be carefully checked in the preflight inspection. A pitot cover is often fitted to prevent water or insects accumulating in the tube when the aircraft is parked. The pitot tube and static vent should not be tested by blowing in them, as they are connected to very sensitive instruments.

A schematic of the pitot-static system can be seen on the following page.

Pitot-Static System Diagram - Conventional

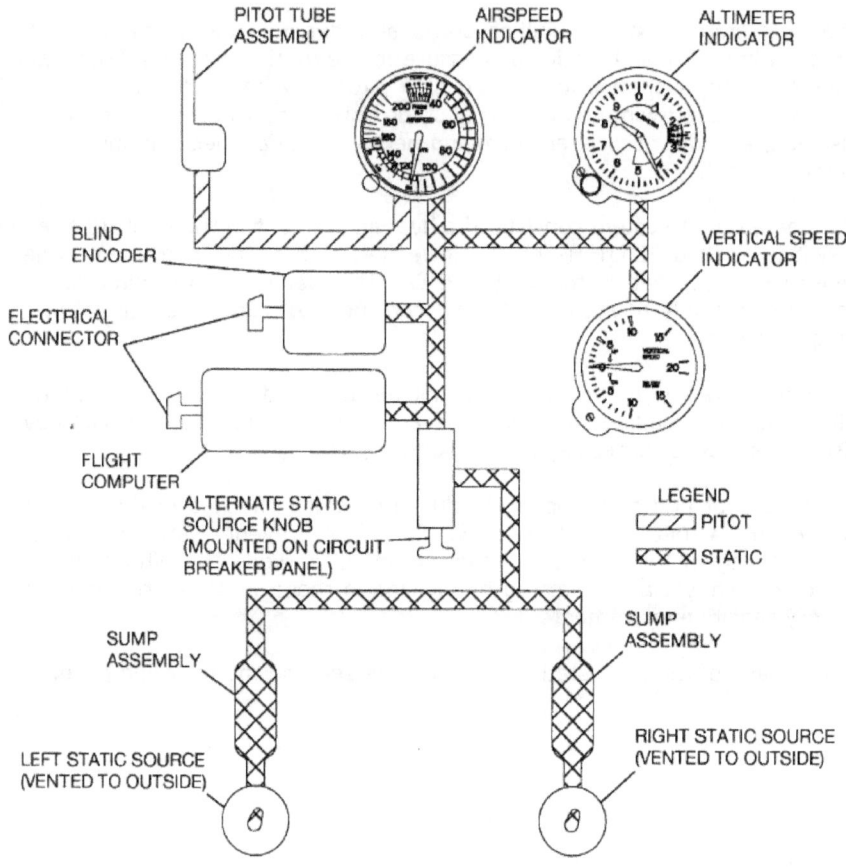

Illustration 14d Pitot-Static Schematic Diagram C206H
(Conventional Instruments)

Pitot-Static System Diagram - Glass

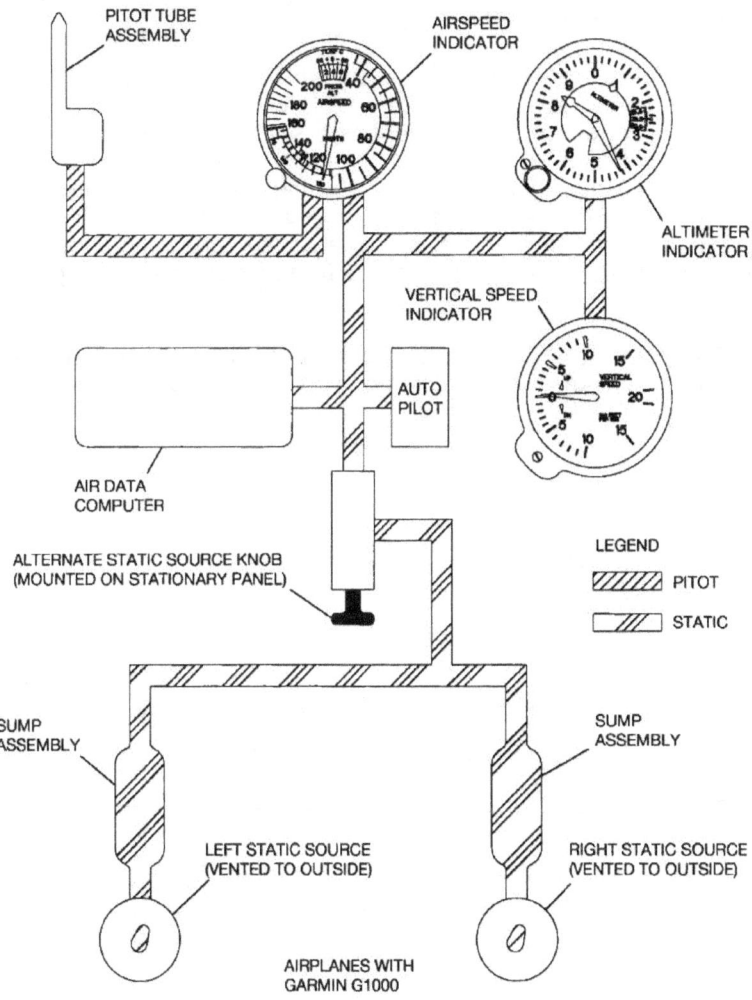

Illustration 14e Pitot-Static Schematic Diagram C206H

(G1000 Instruments)

Vacuum Operated Gyro Instruments

Suction is necessary to operate the main gyro instruments, consisting of the attitude indicator and directional indicator.

Suction is provided by a dry-type, engine-driven, vacuum pump. A suction relief valve, to control system pressure, is connected between the pump inlet and the instruments. A suction gauge is fitted on the instrument panel and indicates suction at the gyros. A suction range of 3 to 5 inches of mercury below atmospheric pressure is acceptable. If the vacuum pressure is too low, the airflow will be reduced, the gyro's rotor will not run at the required speed, and the gyro instruments will be unreliable. If the pressure is too high, the gyro rotor's speed will be too fast and the gyro may be damaged.

Note: The gyro driving the Turn Indicator is normally operated by electrical source to provide redundancy in case of a failure of the vacuum system.

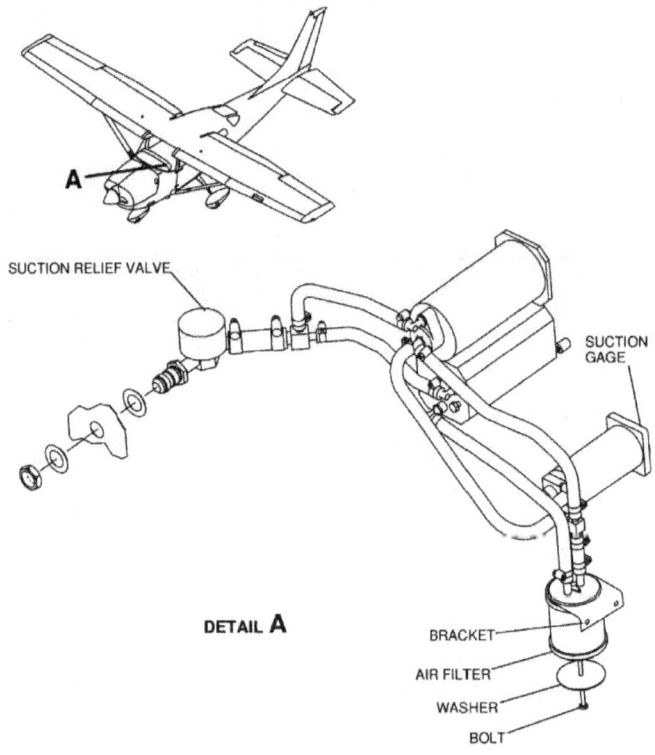

Illustration 14c Suction Pump Installation

Stall Warning

The aeroplane is equipped with a vane-type stall warning system installed on the leading edge of the left wing. The unit is electrically connected to a dual warning unit located above the right cabin door behind the upholstery.
The vane in the wing unit senses the changes in airflow over the wing. When an angle of attack close to the stall is reached, the vane moves up, activating the dual warning unit, which produces a continuous tone over the aircraft's speaker. This typically occurs approximately 5 to 10 knots above the stall in all configurations. Turbulent air may intermittently activate the stall warning at much higher speeds.

The system can be checked during the preflight inspection by turning on the master switch and actuating the vane on the wing leading edge (gently push it up). A sound from the warning horn will confirm that the system is operative.

Illustration 14e Stall Warning

Avionics

The minimum standard fitting is a single VHF radio with hand mike and single jack point. Many aircraft have an additional dual place intercom, and a "Push To Talk" (PTT) button on the pilot's side installed.
Most aircraft will have had upgrades on the avionics systems, so an overview of general operation is included. Explanation of the operation of individual systems can be found in the supplements to the aircraft's Pilot's Operating Handbook.

Audio Selector

Before operation of any radio installation the audio selector panel should be checked. The audio selector selects the position of the transmitter and receiver for the radio equipment on board.

The common selector positions are:

Transmitter: Transmit on COM 1, COM2, PA, Intercom (as installed)

Receiver: Listen to COM1, COM 2, Both, Intercom, Nav ident
Listen to each channel on speaker, head phone or select off

When two radios are installed, it is recommended to develop the habit of always COM1 for the primary active frequency, and COM2 for supplementary information, for example ATIS, company, or the next frequency in unmanned operations.

When there is a need to listen to more than one frequency, be ready to deselect the secondary frequency in case of any important radio transmissions.

Intercom

The intercom, sometimes incorporated in the audio select panel, contains at least a volume and squelch control. The volume control is for the crew speech volume and the squelch for intensity of crew voice activation.

VHF Radio Operations

Once the audio panel has been set, the crew communication established, if required, and the radio switched on, correct operation should be confirmed prior to transmitting. All VHF radio installations will have a squelch selection to check volume and for increased reception when required. This is either in the form of a pull to test button or a rheostat, turned, until activation is heard. Thereafter initial contact should be established if on a manned frequency. Most modern radio installations have an indicator to confirm the transmit button is active. This should be monitored on the first transmission and frequently during initiating radio transmission thereafter.

Illustration 15a Radio Stack

Transponder

Wherever installed transponders should be switched to standby after start to allow for warm up time. When entering an active runway for departure, until leaving the active runway at the end of the flight, the selector should be in ALT if available or ON. Many commercial aircraft now contain TCAS and can observe other transponder equipped targets for traffic separation purposes.

The following ICAO specified transponder codes are useful to remember:

Where no code is specified	2000
Emergencies	7700
Radio failure	7600
Unlawful interference	7500

Ancillary Systems

Lighting

Instrument and control panel lighting is provided by flood lighting, and integral lighting (internally lit equipment) and, optional post lights (individual lights above the instruments).

Two rheostat control knobs on the lower left side of the control panel, labelled PANEL LT and RADIO LT, control intensity of the lighting.

A slide-type switch on the overhead console, labelled PANEL LIGHTS, is used to select flood lighting in the FLOOD position. Flood lighting consists of a single red flood light in the forward part of the overhead console. To use the flood lighting, rotate the PANEL LT rheostat control knob clockwise to the desired intensity.

Individual post lighting may be installed as optional equipment to provide for non glare instrument lighting. The post light consists of a cap and a clear lamp assembly with a tinted lens. The intensity of the instrument post lights is controlled by the instrument light dimming rheostat located on the switch panel.

The external lighting system typically consists of:
- Navigational lights on the wing tips and at the top of the rudder;
- Dual landing/taxi light mounted in the front cowling or wing;
- A rotating beacon located on top of the vertical fin;
- Strobe lights installed on each wing tip;
- Courtesy lights under the wings.

All lights are controlled by switches on the lower left side of the instrument panel. The switches are ON in the up position and OFF in the down position.

The courtesy lights consist of one light located on the underside of each wing to provide ground lighting around the cabin area. The courtesy lights have a clear lens and are controlled by a single slide switch labelled "Utility lights", located on the left REAR door post. Note: Difficulty in finding this switch means pilots often leave these lights on, the switch is not accessible from the cockpit and should be turned off during the pre-flight inspection if not required or after boarding the passengers.

Cabin Heating and Ventilating System

Heated air and outside air are blended in a cabin manifold just aft of the firewall by adjustment of the heat and air controls. The temperature and volume of airflow into the cabin is controlled by the push-pull CABIN HT and CABIN AIR control knobs.

The air is vented into the cabin from outlets in the cabin manifold near the pilot's feet. Wind shield defrost air is also supplied by a duct leading from the manifold to the outlets below the wind shield.

For cabin ventilation, pull the CABIN AIR knob out. To raise the air temperature, pull the CABIN HT knob partially or fully out as required. Additional direct ventilation may be obtained by opening the adjustable ventilators near the upper left and right corners of the wind shield.

The cabin heating system uses warm air from around the engine exhaust. Any leaks in the exhaust system can allow carbon monoxide to enter the cabin.
To minimize the effect of engine fumes, fresh air should always be used in conjunction with cabin heat.

Carbon monoxide is odourless and poisoning will seriously impair human performance, and if not remedied, could be fatal. Pilots should be aware and keep alert to the onset of CO poisoning, especially when operating the cabin heating. Personal CO detectors are inexpensive and available at most pilot shops.

Cabin Heating and Ventilating Schematic

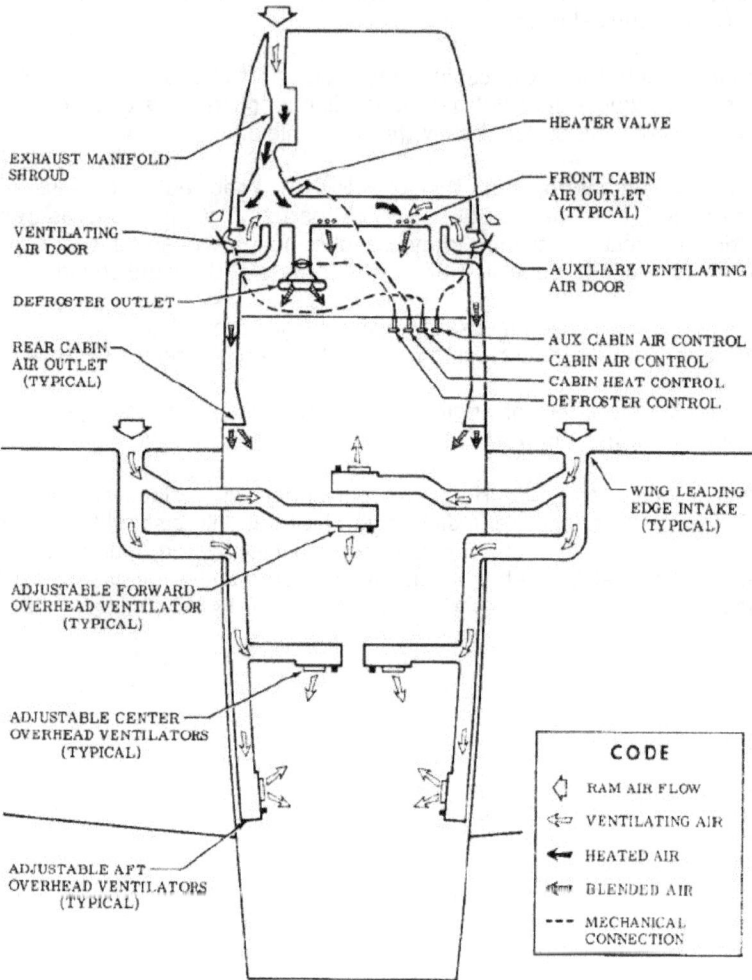

Illustration 16a Cabin Heating And Ventilation Schematic

FLIGHT OPERATIONS

The following sections have been compiled from some of the common Cessna Pilot's Operating Handbooks for the C206 series. Additional notes and handling tips have been included from certified flight instructors and flight engineers. Most speeds have been rounded up to the nearest 5kts.

POH's vary slightly between models, therefore this information is provided as a GUIDELINE ONLY. For operational purposes the POH from the aircraft you are flying, (which by law must be on board the aircraft during flight), should be referred to.

NORMAL FLIGHT PROCEDURES

Pre-flight Inspection

The preflight inspection should be done in anticlockwise direction as indicated in the flight manual, beginning with the interior inspection.

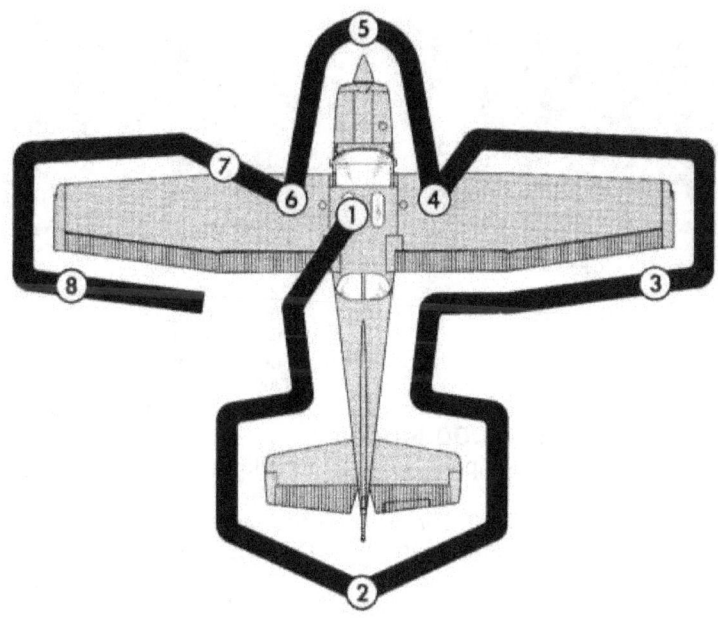

Cabin

Ensure the required documents (certificate of airworthiness, maintenance release, radio licence, weight and balance, flight folio, flight manual, and any other flight specific documents) are on board and valid. Perform a visual inspection of the panel from right to left, and top to bottom to ensure all instruments and equipment are in order.

Control lock – REMOVE
Ignition switch – OFF
Lights - OFF except beacon
Master switch – ON
Fuel quantity – CHECK

Confirm cargo door closed,
Flap lever – DOWN
External Electrical Equipment – CHECK if required (lights, pitot,)
Master switch – OFF
Fuel shut-off valve – APPROPRIATE TANK (the fullest tank or the tank desired to be used for start).

Cabin Inspection G1000 Models

Additionally for G1000 equipped aircraft the following items need to be checked:
With the master switch on:Ensure PFD display visible, check the required annunciators are displayed. Confirm both avionics fans are operational – turn on each of the avionics buses separately and confirm the fan can be heard.
With the master switch OFFTest the standby battery – hold in the TEST position for approx 20 seconds ensure green light remains on. (This test is described before start in the POH, however if the standby bettery is required for flight it is preferable to complete the test now).

Operational Check of Lights

Before turning master switch off, if lights are required, switch all lights on, confirm their operation visually, then turn all off again except beacon. This is required for a night flight and a good idea for all flights.

Exterior Inspection

Visually check the aeroplane for general condition during the walk-around inspection, ensuring all surfaces are sound and no signs of structural damage, worked rivets, missing screws, lock wires or loose connections.

Aft Fuselage, Left Side

Check left static port for blockage.

For passenger versions, once loading is complete, ensure the baggage door is secure. *Caution, worn locks have led to inadvertent opening in flight, where possible bar opening with tow bar or similar device, and lock the door to be sure.*

Check general condition of aft fuselage and windows.
(Cargo version has no rear baggage door on left side).

Tail Section

Check top, bottom, and side surfaces for any damage. Ensure balance weights secure. Remove rudder gust lock and tie-downs if installed.

Ensure Elevator secure and undamaged. Check all linkages free, lock pins in place. Check full and free movement of control. Check trim is undamaged and in neutral position.

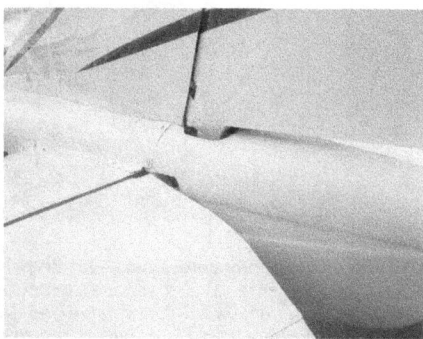 

Rudder linkage and turn-buckles secure and lock wires and pins in place, check for full and free movement of control surfaces.

Check Beacon, rear navigation light, and tail mounted aerials are undamaged and secure.

Aft Fuselage, Right Side and Right Wing

Check the flaps do not retract if pushed, and flap rollers allow small amount of play in down position. Check for damage to surface and flap tracks, operating linkage free movement, adequate grease and security of all nuts and lock pins.
Ensure Cargo Door is closed
Check right side static port (if installed).

Check aileron surface for damage, security of hinge point, and ensure full and free movement. Check wing tip vent unobstructed.

Check condition, security and colour of navigation light.

Ensure all aerials are undamaged and secure. Check top and bottom wing surfaces for any damage or accumulations on wing. *Ice or excessive dirt must be removed before flight.* Remove wing tie down and vent covers if installed.

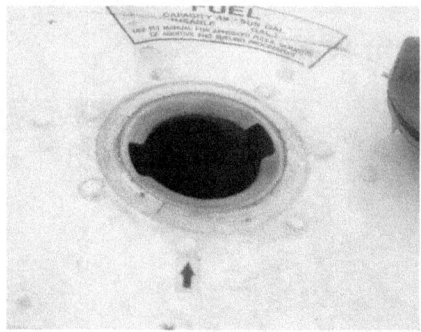 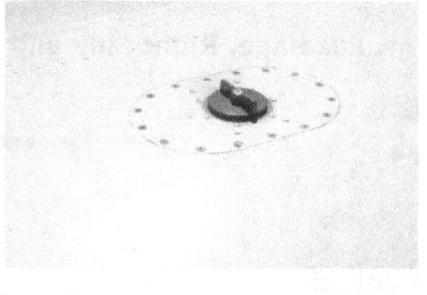

Check visually for desired fuel level using a suitable calibrated dipstick.

Check that fuel cap is secure, and vent is unobstructed.

Check for security, condition of strut and tyre. Check tyre for wear, cuts or abrasions, and slippage. Recommended tyre pressure should be maintained. Remember, that any drop in temperature of air inside a tyre causes a corresponding drop in air pressure and vice versa.

Use sampler cup and drain a small quantity of fuel from tank sump quick-drain valves, under the wing and underneath the cabin, checking for water, sediment and proper fuel grade (first flight of the day and after refuelling). Ensure the drains are seated correctly and not leaking.

Special note regarding precautions during fuelling and fuel drains:
With integral tanks it is possible to have airlocks in the interlinked compartments, resulting in the wing tank appearing full but containing less than full capacity. Whenever maximum range is required, ensure the aircraft is level, and filling is carried out slowly, allowing time to settle. For bladder tanks, it is common for water to be trapped in crevices of the bladders where they become unstuck, and this also requires rocking of the wings after filling to resettle the water in the sump. Fuel should be allowed to settle a minimum of 15 minutes, before draining. Ideally fill up the previous night.

Check security of nuts and split pins, operating linkages, and security and state of shimmy damper. Visually check exhaust for signs of wear, on first flight, if engine is cool check exhaust is secure.

Check oleo for proper inflation and damping. Check cowl flaps for rigidity and operation.

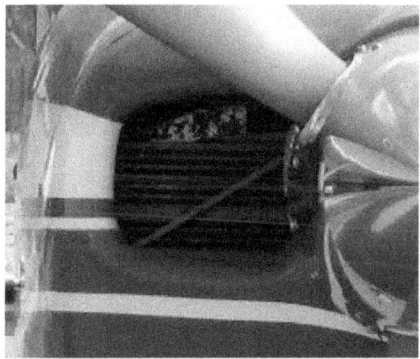

Check oil cooler secure and unobstructed, and alternator belt secure. Ensure no debris inside engine cowls from birds or other sources.

Check condition and cleanliness of landing light, condition and security of oil filter. Check propeller and spinner for nicks and security. Ensure propeller blades and spinner cover is secure. When engine is cold the propeller may be turned through to assist with pre-start lubrication. *Always treat the propeller as live.*

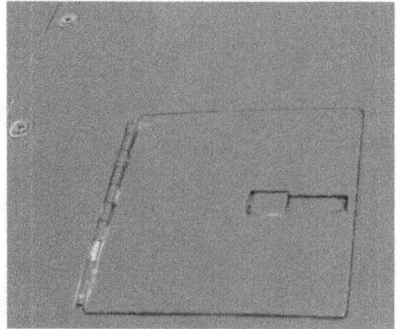 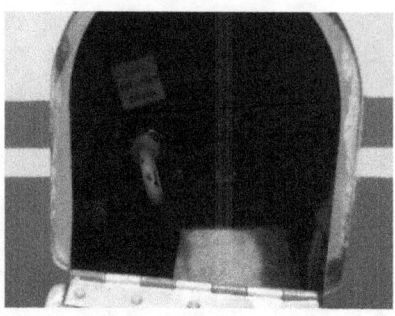

Check oil cap is secure (by opening the top oil filler panel, or through front cowl opening), and ensure oil filler panel is closed securely.

Open inspection cover, check oil level. Minimum oil 7 quarts, fill to 10 for extended flights.
Before first flight of the day and after each refuelling, pull out fuel strainer to check the fuel sample. Check strainer drain is fully closed.

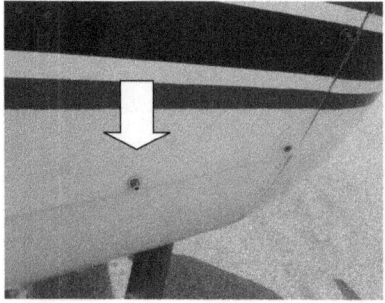

Check security and condition of engine cowling. *On the picture nut on the left is unsecured.*

Differences on the Left Side

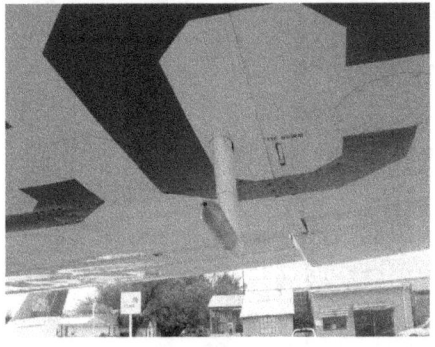 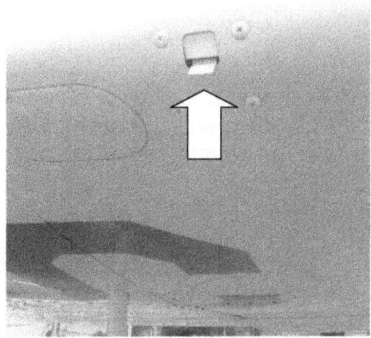

Remove the pitot tube cover, and check the pitot tube for cleanliness, security and clear opening passage.

Check operation of stall warning.

Conduct the check of the left wing in the same manner as the right.
(Except the navigation light should be red!)

Final Inspection

Just before climbing in and starting the engines, complete a final walk around, to save embarrassing, costly or even dangerous oversights.
Check all chocks and covers are removed, fuel/oil caps and door latches are secure, and the aircraft is in a position to safely taxi without excessive manoeuvring or power application.

Passenger Briefing

After completion of the preflight inspection and preferably before boarding the aircraft, take some time to explain to the passengers safety equipment, safety harnesses and seat belts and operation of the doors/windows.

The following items should be included:
+ Location and use of the Fire Extinguisher;
+ Location and use of the Axe;
+ Location of the First Aid Kit;
+ Location of emergency water and normal water;
+ Location of any other emergency equipment;
+ Latching, unlatching and fastening of safety harnesses;
+ When harnesses should be worn, and when they must be worn;
+ Opening, Closing and Locking of doors and windows;
+ Rear door (when installed): emergency opening with flaps extended;
+ Actions in the event of a forced landing or ditching;
+ Cockpit safety procedures (front seat passenger) and critical phases of flight.

Starting

Before engine start or priming is carried out, all controls should be set in the appropriate positions and the panel pre-start scan completed. Priming before you are ready to start can be counter-productive, since by the time you are ready to start the priming fuel is no longer at the point of combustion, but has assisted in washing any residual oil off the cylinder walls.

Checklists before start may be broken down into 'master off' and 'master on' checks, or more correctly named 'before start', and 'ready to start' checks. The latter items are done only once the aircraft has a start clearance, and is in a position to immediately start the engine. The reason for splitting up the checklist is that certain items such as selecting the master on and priming the engine ideally should not be done too far in advance of the start, as the delay will run down the battery and reduce the effectiveness of the priming.

+ Once before start flows are completed, the following before start checks are recommended:
    - **Preflight Inspection** – COMPLETE;
    - **Tach/Hobbs/Time** – RECORDED;
    - **Passenger Briefing** – COMPLETE;
    - **Brakes** – SET/HOLD;
    - **Doors** – CLOSED;
    - **Seats / Seatbelts** – ADJUSTED, LOCKED;
    - **Fuel Selector Valve** – BOTH/CORRECT TANK;
    - **Cowl Flaps** – OPEN;

- **Pitch** – FULL FINE;
- **Magnetos** – BOTH;
- **Avionics** – OFF;
- **Electrical Equipment** – OFF;
- **Rotating Beacon** – ON.

➔ Once ready to start with the master switch ON, complete the 'ready for start' or 'cleared for start' checks:
- **Annunciators** – CHECK (if applicable);
- **Circuit Breakers** – CHECK IN;
- **Mixture** – RICH / AS REQUIRED*;
- **Prime** – AS REQUIRED (50-80lbs);
- **Throttle** – SET approx ½ centimetre**;
- **Propeller Area** – CLEAR.

*For starting purposes the mixture should be full rich at all altitudes. After successful starting, above 3000ft density altitude, leaning is required to prevent spark plug fouling during ground handling at low power settings.
Starting for Lycoming engines (C206H) will require the mixture to be at cut-off until the engine fires. If purging is required before priming, the mixture will also need to be set at cut-off.
**The throttle should be advanced approximately ¼ inch to provide the correct amount of fuel for starting. If the throttle is advanced too much flooding or backfiring can occur, which can lead to an induction fire.
Note: Before engaging the propeller, it is vital to check that the propeller area is clear.

The engine is started by turning the ignition key into START position, to turn over the engine. The key is sprung loaded back to the BOTH and can be released once the engine starts.

On starting, engine RPM should be kept to a minimum until the engine oil pressure has begun rising. If the throttle has been advanced during starting it is important to ensure it is *immediately* reduced as the engine begins to run. In no circumstances should the engine RPM be allowed to over-rev on start up as the oil will not yet have reached all the moving parts.

Once the engine is started and the oil pressure has stabilised, the throttle should be adjusted to idle at approximately 1000rpm.

After starting, if the oil gauge does not begin to show pressure within 30 seconds, the engine should be stopped and the fault reported to the maintenance before any further starts should be attempted. Running an engine without oil pressure can cause serious engine damage.

Priming, Purging and Flooded Starts

A fuel injected engine requires different priming techniques to a carburettor fuel system, and also suffers from starting problems caused by "fuel vapour locks". These items often lead to confusion resulting in failed attempts to start.
The terms and requirements for the normal starting, as well as flooded starts which may occur after over-priming or incorrect purging of the vapour locks, are explained fully in the following paragraphs.

Priming

Priming is achieved by use of the high pressure fuel pump (i.e. selecting both sides of the fuel pump switch). Once ready for start select the fuel pump on and advance the throttle to the desired setting.
When the engine is cold, it must be primed or starting is very difficult. Priming introduces additional fuel into the engine to assist with obtaining the initial combustion to start the engine running cycle.

The amount of priming required to achieve effective starting will depend on the ambient and engine temperatures.
Warm engine starts do not normally require priming, although a small amount, 4-6Gal may assist starting a warm engine.

Priming Lycoming versus Continental

To complicate matters, the different types of engine installations, Lycoming and Continental require slightly different levels of priming for effective starting. Generally speaking a Lycoming engine requires less priming than a Continental. Where a Continental engine starts better with a little priming even when hot and does not flood easily, a Lycoming engine floods easily and normally does not need priming when hot.

Continental

For a cold engine typically full throttle is used and a fuel flow of approximately 8-10Gal/80-130lbs should be noted on the fuel flow gauge. An intermediate throttle setting may be used if less priming is desired, for example on a warm day or after recent operation.

Cessna recommends 8-10Gal priming for a normal start, if unsuccessful repeat the procedure then attempt a flooded start. That is, by deductive reasoning, if you've tried a normal start and you are now pretty sure the engine is over-primed, (*and you've made sure ALL the settings are correct – eg fuel selector*), a flooded start should work.

Lycoming

For the Lycoming engine, Cessna recommends priming only when the engine is cold. Additionally, because of the tendency to flood, they also recommend a procedure similar to a flooded start. A normal start is always commenced with the mixture idle cut off, feeding the mixture in as the engine fires. In this case the throttle is set for idle (rather than full open as in the flooded start), making the procedure slightly easier to manage. See more below on starting the C206H

Purging Fuel Vapour

Purging is required to clear the vapour locks that fuel injected engines frequently suffer from. The term 'vapour lock' refers to a situation where the fuel in the lines has vaporised and the trapped gasses cause a blockage in the fuel lines preventing adequate supply of fuel getting to the engine. The vaporising is caused either by engine heat or high outside temperatures and if it is not cleared it can make it extremely difficult to start.

To clear or 'purge' fuel vapour locks, high pressure fuel is cycled through the fuel lines by using the high pressure fuel from the auxiliary pump, with the mixture in the idle cut off position.

The important difference between priming and purging is that in purging, fuel does not go to the engine. Therefore, provided the mixture is at idle cut off there is no danger of 'over-purging' unless you have a weak battery. The fuel flow indicator should show no fuel flow, as the indicator displays the fuel flow to the engine, downstream of the mixture control.

If fuel vapour persists during starting, the engine may run for a few seconds then die. If this occurs select the auxiliary fuel pump momentarily on high to clear the vapour and prevent the engine stopping. The pump should be operated for approximately 1 second as the engine starts to decrease, operation for longer periods may cause the engine to cut due to flooding. Selection of the fuel pump on low will also assist if the vaporisation is less severe, for example the engine is running but unevenly. The fuel pump should be selected off again once the engine has stabilised and is running smoothly at idle.

Flooded Starts

Engine 'flooding' is caused by over priming, and means there is excess fuel in the engine. The excess fuel will make starting very difficult as the mixture is much less combustible and ignition may be hampered by wet plugs, thus an engine clearing procedure is required.

To remove the excess fuel from the engine, the engine is turned over with the mixture at idle cut-off. As the engine is turned over without fuel supply, the air entering the cylinder during the cycle clears out the excess fuel. To ensure maximum air flow the throttle should be fully open. With the Mixture Idle-Cut-Off no fuel will be permitted to enter the engine, however always cross check to ensure the mixture is completely closed by checking the fuel flow gauge. The fuel selector may be selected off if it is suspected there is fuel leaking through at Idle-Cut-Off position.

Flooded Starting combines engine clearing with engine starting. With the mixture control at idle-cut-off, the throttle fully open*, and fuel selector ON, the fuel lines should be purged to ensure fuel is in the lines (Fuel pump selected on for 10-15 seconds). Thereafter motoring is initiated, as with the engine clearing procedure. Because of the excess fuel in the engine, the mixture in the cylinders will begin to ignite, when this happens the mixture is advanced and the throttle retarded to maintain idle RPM.

*Half open is recommended in the C206H.
Generally when motoring the engine it is difficult to tell when you might reach a combustible mixture so the two procedures (clearing and starting) should normally be completed simultaneously. When clearing only is desired, as there is no way to isolate the ignition, the fuel selector should be selected to off.

If continuous cranking is required, ensure starter limits, not more than 30 seconds without cooling, are observed.

Pre-Heat

When operating in extreme cold temperatures, preheat will be required to thaw and de-congeal the oil to provide adequate lubrication.
Use of an external power source is recommended wherever possible to prevent strain on the electrical system.

Full details of operation of the preheat are contained within the supplements section of the Pilot's Operating Handbook.

Starting Procedure

The engine is started by turning the ignition key into START position. Before engaging the starter ensure the area is clear, be looking outside and keep one hand on the throttle for adjustment during starting or as the engine fires, and feet on the brakes (light aircraft park brakes are not always self adjusting and the park brake may have become weak with brake wear). In the case of the Lycoming or flooded start procedure, initially your free hand will be on the mixture, feeding it in slowly as the engine fires, however return immediately to the throttle to prevent any over-revving.

Do not crank the engine continuously if the engine fails to start. The starter motor should not be operated continuously for more than 30 seconds. Additionally, if the engine fails to start, typically it is either under or over primed, or you have omitted an important step, eg placing the fuel selector on, and a quick review of the procedure and settings should be completed before another attempt.

The following section details the manufacturers recommended procedure for starting the most common engine configurations for the C206. Refer to the POH from the aircraft you are flying to confirm if there are any differences.

Starting the C206G and Earlier models

The recommended (pilot's operating handbook) start procedure for a Continental fuel injected engine is as follows:

Starting Engine (With Battery) – C206G and Earlier
1. Throttle – FULL
2. Propeller – HIGH RPM
3. Mixture – RICH
4. Auxiliary Fuel Pump – HIGH 3-4 SECONDS AS REQUIRED then OFF
5. Throttle - ¼ INCH
6. Propeller Area – CLEAR
7. Master Switch – ON
8. Ignition Switch – START (release when engine starts)

Starting the C206H

The latest model Cessna 206, with the larger fuel injected Lycoming engine, is more sensitive to over-priming. Because of this Cessna recommends starting with the Mixture Idle Cut Off, then advancing as the engine fires, as detailed below (note; this procedure is similar to the flooded start procedure, however the throttle remains at ground idle setting). Priming is carried out using the fuel pump, normally only when the engine is cold.

Starting Engine (With Battery) – C206H
1. Throttle – OPEN ¼ INCH
2. Propeller – HIGH RPM
3. Mixture – IDLE CUT OFF
4. Propeller Area – CLEAR
5. Master Switch – ON
6. Auxiliary Fuel Pump Switch – ON
7. Mixture – FULL RICH - 3 to 4 seconds fuel flow, then return to IDLE CUT OFF position
NOTE: If engine is warm, omit priming procedure of step 7 above

8. Ignition Switch – START (release when engine starts)

9. Mixture – ADVANCE smoothly to RICH when engine fires
In both cases, as engine starts, the ignition switch should be released into the BOTH position and the throttle adjusted to 1000 engine rpm or less. In no circumstances should the engine be allowed to over-rev on start up, as damage may occur from applying excessive rpm when the engine has had insufficient time to lubricate.

After starting, if the oil gauge does not begin to show pressure within 30 seconds, the engine should be stopped and reported to the maintenance. Lack of oil pressure can cause serious engine damage.
Selection of the fuel pump to low after start may be required to prevent fuel starvation due to vaporisation in high ambient temperatures. The fuel pump can be left on until the engine rpm stabilises at idle.

After Start

After start flows are completed, and should be followed by an after start checklist, to ensure all the critical items are completed prior to taxi.

At airfields above 3000ft density altitude, the mixture should be leaned for taxi to prevent spark plug fouling. When operating at airfields which require full rich mixture for takeoff, it is best not to lean the mixture for taxi, lest you forget to enrich it again, resulting in a potentially tragic situation of partial power or an engine cut on departure.

A "live mag" check should be done at this point, by selection of the left and right positions to confirm both are operating. This is not an integrity check as the engine is still cold. The purpose of the check is to prevent unnecessary taxiing to the run-up point should one magneto have failed completely.

Before taxi, the direction indicator must be set to the compass for orientation purposes, and the transponder set to standby for warm up, so that it is ready for use on departure.

✈ The following after start checks are recommended:
- **Mixture** – SET for taxi;
- **Magnetos** – CHECKED;
- **Engine Instruments** – CHECKED;
- **Flaps** – RETRACTED/SET;
- **Transponder** – STANDBY/GROUND.

The time spent completing the after takeoff checks properly will also assist with the engine warm-up prior to taxi.

Warm Up

If the engine is cold, for example on first flight of the day, or when it is anticipated that high power may be needed during taxi, time should be allowed for the engine to warm up prior to taxi. Ideally this warm up period should be sufficient to allow the CHT to increase into the green range before taxi. The cowl flaps should not be closed for this warm up as this will provide uneven temperature distribution which may damage the engine.

Most of the warm-up needed prior to completing the engine run-up can be accomplished during taxi. If the flight is from an airfield where only a very short taxi is required, additional warm-up time should be allowed before engine run-up or take-off is carried out.

Taxi

Taxi speed should be limited to a brisk walk, the aircraft is at its most unstable condition on the ground, especially with strong winds that may reach minimum flying speeds.

Controls must be held to prevent buffeting by the wind. The elevator should be held fully aft when taxiing over rough surfaces, bumps or gravel to

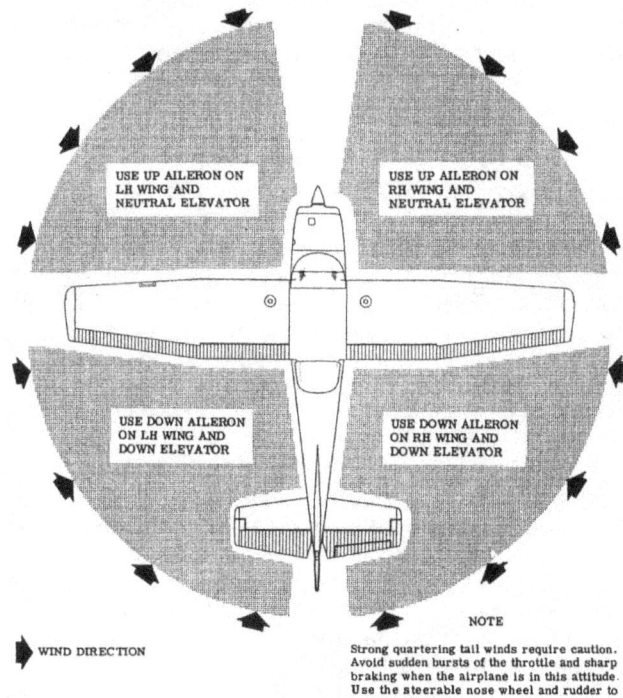

NOTE

Strong quartering tail winds require caution. Avoid sudden bursts of the throttle and sharp braking when the airplane is in this attitude. Use the steerable nose wheel and rudder to maintain direction.

reduce loads on the nose wheel and propeller damage. In all other cases the diagram following illustrates positions of controls in consideration of wind for the best aerodynamic effects during taxi.

The following phrase may be helpful as a memory aid:
CLIMB INTO WIND
DIVE AWAY FROM THE WIND

Brake use should be kept to a minimum by anticipation of slowing down or stopping followed by reduction of power to idle prior to applying brakes. This ensures brake wear is minimised and ensures that brakes are still at their most effective should they be required for an aborted takeoff (excessive use will increase brake temperatures and reduce their effectiveness). Brakes should NEVER be used with power on.

Engine Run-up

The engine run-up is usually performed at the holding point, in a clear area to avoid disturbances to others on the manoeuvring areas. The larger engine on the C206 creates significant noise and prop-wash so care should be taken in positioning.

Before completing the run up it is important to ensure the fuel is on the correct tank for take-off, the prop area (ahead and behind) is clear, the mixture is set for higher power operations enrich slightly if leaned for taxi), and the temperatures and pressure are within the acceptable limits for the run-up.

The engine run-up is completed at 1700rpm.

Advance the engine to 1700 rpm and perform the following checks:
+ Prior to take-off from fields above 3000ft density altitude, the mixture should be leaned. As the air pressure decreases with altitude the air density also decreases and so the engine receives less mass of air. If the mixture is left in the full rich position, the air/fuel ratio will not be correct (too much fuel or the mixture too rich). The correct air/fuel ratio is required for engine to produce maximum available power.
+ The following procedure may be used for leaning the mixture during the run-up: lean the mixture by rotating the mixture knob anticlockwise till peak RPM, then enrich the mixture for about 3 rotations. This procedure is similar to that carried out enroute for leaning. This check may also be performed at lower altitudes to check correct operation and setting of the mixture. The mixture should be increased approximately half the amount it has been leaned, or below 3000ft returned to full rich, before takeoff;

↛ The propeller CSU should be checked to confirm correct operation and to ensure proper lubrication throughout the governor.
  - Select the pitch control to full course, noting:
  - RPM drop, (visually and audibly);
  - manifold pressure increase;
  - oil pressure drop;
  - return to fine again to prevent RPM dropping more than approximately 300rpm.
  If the engine is cold repeat the process until the RPM drops smoothly and rapidly, three cycles are recommended.

↛ The Magneto check should be done as follows.
  - Move ignition switch first to L and note RPM.
  - Next move the switch back to BOTH to clear the other set of plugs and regain the reference RPM.
  - Then move the switch to R position, note RPM and return the switch to BOTH position.
  - RPM drop in either L or R position should not exceed 150rpm and show no greater than 50rpm differential between magnetos;

↛ Verify proper operation of the alternator, the suction system, and the correct indications (in the green range) of all engine control gauges;

↛ The DI may be set to compass at this point as the suction pressure and effect of engine interference provide a more accurate setting at 1700rpm;

↛ Reduce the engine RPM to idle to confirm the correct idle setting, now that the engine is warm and the mixture set correctly,

↛ Return to normal idle at 1000rpm, for completion of the pre-takeoff checks and while awaiting take-off clearance.

(Note – the above list is a recommended minimum requirement, some operations may require additional checks).

Pre-Takeoff Vital Actions

The flight manual provides the "minimum required actions" before takeoff, generally there are some additional operational items which have become standard practice to check. Many flight schools or operators will have their own check lists and/or acronyms for the pre take-off checks. Acronyms are highly recommended for single pilot operations, especially for critical checks such as pre-takeoff and pre-landing.

One of the more popular pre-takeoff acronyms is detailed below.

Too   Trims and controls and tested and set;
Many Mixture set for takeoff;,
         Magnetos on both;
Pilots Pitch full fine;

Go	Pumps on (as applicable); Gills open; Gyros uncaged and set (as applicable);
Fly	Fuel contents checked on correct tank, primer locked, pump as required (normally off for the C206); Flaps set for takeoff (recommended 10 degrees);
In	Instruments, check panel from right to left, DI aligned with compass, navigation and radio aids set, time check;
Heaven	Hatches and harnesses secure;
Early	Electrics charging (verify with landing light or flap if desired), circuit breakers checked, systems set.

Caution the inherent trap in this checklist of having two different items for a number of letters. Following the memory checks with a written checklist is the best way to avoid errors, and should be considered mandatory, especially for demanding flight situations such as single pilot IFR.

Line-Up Checks

Line up checks are completed once lined up on the runway. If a rolling take-off is necessary due to traffic or runway conditions, line-up checks are completed during the line up.

A useful acronym for line up checks is 'Remember What To Do Last':

REmember	Runway unobstructed, correct, and nosewheel aligned Engine temperatures and pressures check
What	Windsock check wind direction and strength, (confirm against ATC wind), position control column accordingly;
To	Transponder selected to ALT mode (TA/RA or ON as applicable);
Do	DI aligned with compass and indicating correct runway;
Last	Landing light and strobes – ON.

Takeoff

Takeoff is carried out under full power with the heels on the floor to avoid accidental application of the toe brakes.

It is important to check correct engine operation early in the takeoff run. Any sign of rough engine operation or sluggish engine acceleration or less than expected takeoff power is cause to discontinue the takeoff. The engine should run smoothly and with constant maximum static RPM (red line), manifold indicating within 1-2 inches of ambient pressure (normally aspirated) or maximum (turbo), and fuel flow checked according to the maximum power fuel flow placard.

When taking off from gravel runways, wherever possible the throttle should be advanced slowly to allow the aeroplane to start rolling before high RPM is developed, to minimise damage from the loose gravel on propeller. In a rolling takeoff the gravel will be blown back from the propeller rather than pulled into it.
In this situation takeoff length is increased by the distance it takes to reach full power.

Fuel flow Setting for Takeoff

Fuel flow setting must be corrected using MIXTURE control early in the takeoff roll. If it was properly set during engine run-up, it will only require minor changes. It is obvious that, to achieve this, the fuel-flow placard setting should be noted BEFORE and not during take-off run. For this reason, the take-off crew briefing or self-briefing should always include the power setting (RPM, manifold pressure, *and* fuel flow) required for takeoff, to ensure a timely abort is initiated if less than maximum power is available.

An example of a fuel flow placard from the C206H is displayed opposite. The fuel flow placard must be displayed clearly in the cockpit for the model concerned and in the same units as the fuel flow gauge. Fuel flow settings are the minimum the required.

✪ Fuel flows and units vary between models.

MAX. POWER FUEL FLOW	
ALTITUDE	FUEL FLOW
S.L.	28.0 GPH
2000 Feet	26.5 GPH
4000 Feet	25.0 GPH
6000 Feet	23.0 GPH
8000 Feet	21.5 GPH
10,000 Feet	20.0 GPH
12,000 Feet	18.5 GPH
14,000 Feet	17.0 GPH

It should be remembered, that with a turbo charged engine, the red line limitation can sometimes be reached or even exceeded before reaching full-throttle, and so care should be taken when applying full power. A turbo charged engine will also have a fixed fuel flow up to a significantly higher altitude, for example the 206H turbo engine requires 39GPH minimum at 39" Hg (maximum continuous power) up to 17,000ft.

Wing Flap Setting on Takeoff

Normal takeoff may be accomplished with wing flaps set from 0 to 20 degrees. Flap settings greater than 20° are not approved for take off.

Use of flap for takeoff will shorten ground roll but will reduce climb performance of aircraft.

During testing, it is established which flap settings will be most favourable, for the entire takeoff, including the takeoff ground roll and climb to the 'barrier height' of 50ft. The associated performance is tabulated and provided in the performance section with the applicable requirements.

The recommended flap setting for a short field takeoff, to obtain the minimum total takeoff distance is 20 degrees.

The higher the flap setting the lower the speed an aircraft will become airborne, and the additional lift generated throughout the ground roll will shorten the ground roll considerably.
Use of 20 degrees flap on soft or rough surfaces will assist with reducing the frictional drag considerably.

If there is rising terrain after the 50ft point, selection of 0 degrees flap for climb performance, should only be considered when runway length far exceeds that required, as no performance figures are provided for a flap up take-off.

Flaps should not be retracted before reaching a safe altitude of approximately 300ft AGL, and not before the minimum safe flap retraction speed, of 80kts/90mph is reached. With flap retraction the aircraft loses lift, before the effect of the reduced drag is felt, and the stall speed is increased due to the change in wing profile. If retracted at too low speed, significantly loss in climb performance will result, or if the climb performance/angle is attempted to be maintained the aircraft will loose speed, putting it dangerously close to the stall and on the wrong side of the drag curve.

Normal Takeoff

➢ The normal takeoff procedure specified in the Pilot's Operating Handbook is:
- **Wing Flaps** - 0-20 degrees ;
- **Power** – FULL THROTTLE, 2850 (maximum) rpm;
- **Mixture** – LEAN as per placard;
- **Elevator** - LIFT NOSE WHEEL at 50kts;
- **Climb Speed** - Maintain 70-80kts / 80-90mph until obstacles are cleared;
- **Wing Flaps** - RETRACT once obstacles are cleared, and after safe retraction speed of 80kts/90mph is reached.

Short Field Takeoff

For the minimum takeoff distance to clear a 50ft obstacle, the short field or maximum performance takeoff technique should be applied

➢ The recommended procedure in the Pilot's Operating Handbook for a short field take-off is:
- **Brakes** – APPLY*;

- **Wing Flaps** 20 degrees;
- **Power** – FULL THROTTLE, 2850 (maximum) rpm**;
- **Mixture** – LEAN as per placard;
- **Brakes** – RELEASE;
- **Elevator** – SLIGHTLY TAIL LOW, lift off early;
- **Climb Speed** – Maintain 65kts / 75mph until obstacles are cleared, (once obstacles are cleared, accelerate to best rate of climb, Vy);
- **Wing Flaps** – RETRACT once obstacles are cleared, and after safe retraction speed of 80kts/90mph is reached.

* If power is applied after brake release increase distance by the distance taken to apply full power.
**Or the applicable maximum permissible takeoff rpm for the engine installation.

Although not specified in the POH procedure, prior to brake release, it is important that the manifold pressure is confirmed acceptable, not more than maximum 2" below ambient (pre start indicated) pressure or QFE in inches.

There is no speed specified for lift off in the short field or normal takeoff procedure in the POH. With a tail low attitude the aircraft will become airborne as it gains flying speed. This technique is recommended as the sooner the aircraft is airborne frictional drag is removed, however the aircraft needs to be accelerated to the minimum climb out speed (65kts) to overcome the high induced drag at low speed before it will climb away.

If the aircraft is loaded with an aft centre of gravity limit, or when taking off from an uneven runway, it may become airborne well before the recommended lift off speed. This is very dangerous situation as the aircraft will fail to overcome the induced drag, maintain a very slow speed (well below the minimum flap retraction speed) and fail to climb. To overcome this, the aircraft needs to be accelerated in 'ground effect' until sufficient speed is regained to enable safe climb out. If insufficient clearway is available for acceleration, this method will lead to an inadvertent impact with terrain, therefore it is very important to guard against early lift off.

Raising the flaps too early in an attempt to improve the climb performance, at speeds below the minimum recommended flap retraction speed, will likewise lead to loss of lift, resulting in loss of height, and or an inadvertent stall. Both of which may be unrecoverable at low altitude, as many have sadly demonstrated.

The figures and procedure above, are those prescribed in the flight manual for the maximum performance takeoff at maximum weight. Any deviation from the recommended procedure should be expected to give a decrease in performance.

Soft Field Takeoff

Soft or rough field takeoffs are performed with 20 degrees wing flaps by lifting the aeroplane off the ground as soon as practical in a slightly tail-low attitude. If no obstacles are ahead, the aeroplane should be levelled off immediately to accelerate to Vy (best rate of climb) for best initial climb performance. If there are obstacles, the aircraft should be accelerated to Vx (best angle of climb) and this speed should be maintained until all obstacles are cleared.

Crosswind Component

Early models did not specify a crosswind limit, however in later models a maximum demonstrated crosswind component of 15 knots was introduced.

This is the highest value for which the aeroplane handling has been tested satisfactorily during takeoff and landings, but is not considered to be the limiting crosswind velocity.
It is good operating practice to not exceed this limitation during normal operations, and it is also vital that an inexperienced pilot should reduce this value even further. However situations may arise where a landing with strong crosswind is unavoidable, therefore thorough dual practice during conversion training should be completed.

During a crosswind takeoff, as the aircraft becomes airborne, it will tend to move sideways with the air mass and sink back onto ground with strong sideways movement which may damage the undercarriage.
Therefore, the recommended technique is to hold the aeroplane firmly on the ground to slightly higher lift-off speed and then positively lift-off with a backward movement of the control column. Once airborne the aircraft nose is turned into wind to prevent drift, using a very shallow coordinated turn ('weathercocking'). This drift prevention is commonly termed 'crabbing into wind' because of the apparent sideways motion of the aircraft.

Takeoff Profile

Normal takeoff should consist of the actions depicted below in each phase of departure.

Flap, power and speed need to be concisely managed, and there is a specific requirement and order for each at each phase in the takeoff. This profile is generally consistent with all conventional aircraft, since the aerodynamic principles do not change, only the heights and type of controls/terminology will change.

The takeoff profile can be summarised as follows:

1. Minimum speed/recommended rotate speed (approximately 50kts for a normal takeoff): Rotate- raise the nose wheel/lift off, tap the brakes to stop the wheels moving, reducing the vibrations often felt from imbalances when they are allowed to decelerate on their own.

2. At the end of the runway, a minimum speed of 65kts should have been achieved.

3. Once airborne: Accelerate to initial climb speed (65-80kts), best angle of climb (approximately 65kts) when obstacles exist or best rate of climb (approximately 80kts) to achieve maximum height in minimum time and reduce the risk exposure close to the ground.

4. At a safe height away from the ground and above obstacles in the takeoff path: (allowing for further acceleration if required, typically not below 300ft AGL), accelerate to minimum flap retraction (80kts) and raise the flaps.

5. Once flap is retracted, above approximately 500ft AGL, reduce to climb power (maximum continuous). This is done only after you have removed all the drag, and above an altitude permitting a reasonable chance of a safe outcome from an engine failure, whilst observing the take-off power limitation time (typically 5 minutes if applicable). This should be done by first reducing the manifold pressure, then RPM, followed by mixture setting if applicable. Reducing RPM will increase the manifold pressure slightly. Fine-tuning of the manifold pressure may be needed after adjusting the mixture, once all the engine parameters are stable.

6. Continue to climb at best rate of climb until above 1000ft AGL (VMC) or MSA (IMC or in mountainous terrain).

7. Transition to enroute climb: Accelerate to the desired climb profile (90-110kts or approximately 500 ft/min),

8. Complete the after takeoff checks (flows) and/or after takeoff checklist as available.

A takeoff profile summary diagram can be seen below.

Takeoff Profile Diagram

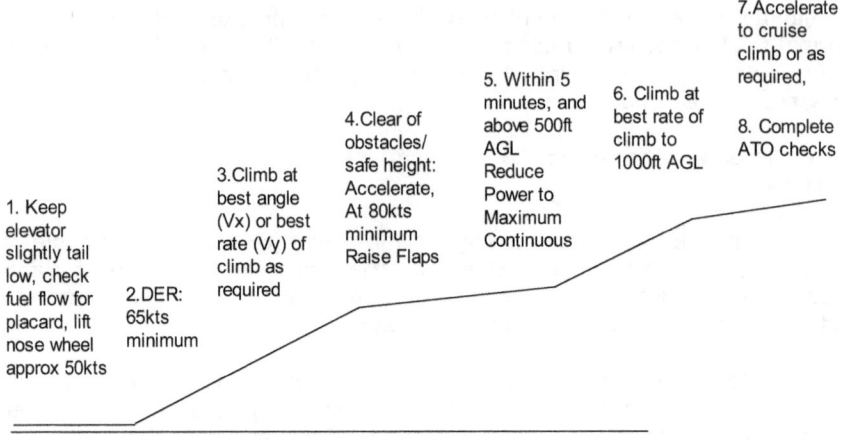

After Takeoff Checks

Typical after takeoff checks are as follows **(BUPPMFFEL)**:
- **Brakes** – CHECK -on and off;
- **Undercarriage** – FIXED;
- **Power / Pitch / Mixture** – SET for climb;
- **Flaps** – UP;
- **Fuel** – CORRECT TANK, CHECKED (LEFT/RIGHT or BOTH as installed);
- **Engine's Temperature & Pressure** – CHECKED;
- **Landing Light** – OFF / AS REQUIRED.

Climb

The normal climb with or without flap, is made at 25" manifold pressure and 2550rpm (the top of the green – normal operating- arcs).

➤   If a maximum rate of climb is desired an airspeed of:
- 84kts/ (97mph) at sea level reduced to;
- 78kts / (90mph) at 10,000ft (approx reduction 2kts/5000ft).

For a maximum performance climb, the maximum permissible power should be used (maximum manifold pressure and 2850rpm). For engine handling considerations maximum power should only be used when needed, and for a maximum of 5 minutes, reducing thereafter to normal climb power.

The maximum rate of climb speed or Vy is used to reach cruise altitude as quickly as possible, as it gains the greatest altitude in a given time.

Normally this speed is only used until safely away from the ground (500 to 1500ft AGL).

When an obstacle clearance climb is required, a maximum angle climb - Vx is used. Best angle of climb gains the greatest altitude for a given horizontal distance.

✈   The best angle of climb speed is:
- 66kts/ (76mph) at sea level;
- 70kts/ (81mph) at 10,000ft.

After takeoff with 20 degrees of flap 65kts is the recommended speed for obstacle clearance.

Because the slow airspeed results in reduced cooling and higher engine temperatures, and reduces the margin above the stall, this speed should be used only when necessary for climb performance, or for practice during dual training.
If sufficient performance allows, a cruise climb at lower altitudes may be achieved by lowering the nose to maintain a rate of climb of 500ft/min. This should provide a speed between 95 to 105kts, (110-120mph). This provides better engine cooling, improves forward visibility, and provides added passenger comfort.
It should be noted that the Cessna 206 requires some degree of mixture management during climb.
Cruising climbs should be conducted at 108 lbs/hrs (which is slightly higher than top of green range 102 lbs/hr) up to 4000 feet. If take off was performed out of an airfield lower than 4000 feet, when reducing to climb power the mixture is reset to 108 lbs/hrs. During the climb as the manifold pressure drops, to maintain climb power, the throttle is constantly increased to return the manifold pressure to the top of green range, thus also increasing fuel flow. As this occurs the mixture is adjusted by reducing fuel flow back to 108 lbs/hrs.

Climbs to higher altitudes than 4 000 feet shall be conducted at the fuel flow shown in the POH Performance Section in the TIME, FUEL, AND DISTANCE TO CLIMB table. Normally, at around 4 000 feet, full throttle setting will be achieved. At this height climbing higher will cause the manifold pressure to drop, which cannot be adjusted as the throttle is fully forward. Despite reaching full throttle height, the mixture will still require adjustment during the climb to maintain a steady climb EGT.

✈   Approximate fuel flow settings for the climb at 2700rpm are:
- S.L to 4 000 – 108 lbs/hr - 23GPH;
- 8 000 – 96 lbs/hr – 19GPH;
- 12 000 - 84 lbs/hr - 17GPH.

These setting may vary significantly between aircraft, even with the same model designator, due to different engine installations. Many models require a climb power fuel flow placard, (as well as a maximum power fuel flow placard) which will reflect the correct minimum fuel flow for the climb. Where this is not placarded, the information is contained in the performance and/or normal operating section of the POH. For this and many other reasons, it is important to remember to review relevant sections in the POH when operating a different C206 for the first time.

Many instructors teach students to set climb power to the top of green (102 lbs/hr) during climb. However this is only accurate at approximately 5-6000ft. Setting 102 lbs/hr when taking off from sea level airfield will cause engine overheating. It should be remembered that excessive fuel is used to cool engine down during the climb to counteract the effects of the reduced airflow at high power settings. Wherever there is doubt, always refer to the manual in the aircraft you are flying for correct settings.

Cruise

Optimum cruise is achieved at the recommended cruise power settings provided in the performance section (see ground planning and performance).
Normal cruise power may be selected in the normal operating (green) arcs for manifold and rpm at pilot discretion. This is typically between 2200 - 2450rpm and 15-25" of manifold pressure (or 22-29" for a turbo engine). For most normal flights a setting of 23" and 2400rpm provides for efficient cruise speed and fuel consumption.
It is often best to set the RPM in cruise at the RPM used for balancing the propeller, this is often decided by the individual maintenance supplier but will be around 2400rpm.

Some engines prohibit certain rpm/manifold combinations, for example the Bonaire Engine is limited to 20" manifold pressure with the rpm below 2200, and may not be operated between 1950 and 2150. Regardless of the engine, too low rpm with too a high manifold pressure is not a good combination.

As altitude increases, on a normally aspirated engine, the manifold pressure will continue to drop and must be increased until "full throttle height" is reached.

To achieve the best fuel consumption, the mixture should be leaned during the cruise. The recommended method is to lean until a loss of manifold pressure and rough running is experienced, thereafter return to slightly rich of peak. With an accurate EGT the temperature should be returned to 50 degrees rich of peak. Operation at peak or lean of peak may cause excessive engine wear through detonation and high operating temperatures. Operating with an excessively rich mixture will lead to spark plug fouling and excessive fuel consumption.

If an aircraft is equipped with individual cylinder EGT and CHT monitoring, the manufacturer of these engine gauges may have a procedure for mixture setting and monitoring. Many installations of this type permit operation at peak EGT, however this must be done with considerable caution and careful monitoring, as a change in ambient conditions may put the mixture towards the lean side of peak, risking detonation or a loss of power. This procedure will be detailed in the associated POH supplement and should be reviewed carefully prior to flight.

Fuel balance should always be monitored during flight, even when operating on BOTH, as there often can be uneven feeding between tanks.
When operating models without BOTH, it is recommended to either change tanks at a standard time interval, eg every 30 minutes, or to change at specified points on your flight log. On short flights (1 hour and less), it is often best to use one tank for takeoff and landing, and another for the cruise, e.g in a one hour flight, takeoff and climb on one tank, change to the other at top of climb for approximately 30 minutes (depending on the length of climb) and return to the first tank for landing. Picking way-points at relatively equal intervals is another great way, as you can indicate a reminder on your flight log, recording the ATA or updating the ETA will remind you about the required change.

During the cruise it is important to have periodic aircraft status checks. These checks will not form part of a checklist, as they are considered normal flying duties and should be done regularly as part of good airmanship, however it is helpful to have an acronym to remind us what to check.

✈   One of the recommended cruise checks is defined by the acronym 'SAFDIE', as follows:
- **S – Suction** – CHECK;
- **A – Amps** – CHECK;
- **F – Fuel** – CHECK sufficient quantity and balanced;
- **D – DI** – ALIGNED with the compass;
- **I – Icing** – CHECK (carb heat/airframe);
- **E – Engine** – CHECK (temperature and pressure, mixture).

Descent

Descent planning should be done with two factors in mind. Speed management, and engine management. Both can result in costly maintenance if a descent is not planned properly.
The most important part of descent planning is to allow sufficient time. Too little time may result in an over-speed during the descent, which can cause

structural damage, or require large reductions in power to achieve the desired speed, resulting in possible shock cooling.

A good rule of thumb is to allow at least 5 miles or 3 minutes per 1000ft *and* per 1" of manifold pressure. For a turbo model these distances can be increased slightly, eg. 6nm per 1000ft etc. Shallow descents and gradual power decreases will not only improve engine handling and prevent over-speeds but they will improve passenger comfort, and ensure you arrive in the circuit at an appropriate speed to configure for the approach.

Descent checks can be completed as memory checks or in a flow pattern followed by a descent check-list, as available. The type of descent checks required may vary depending on the flight undertaken.

The following checks describe a good acronym to encompass both IFR and VFR flight, to be carried out prior to or during the descent.

→ One example of typical descent checks is Tripple A HATFIRE:
- **A – ATIS** – RECEIVED/weather checked;
- **A – Aids** – TUNED (Navigation and Approach Aids) ;
- **A – Approach** – BRIEFED;
-
- **H – Heading** – CHECKED, heading aligned/synced, track/wind noted;
- **A – Altitude** – CHECKED, descent profile check, MSA check, QNH set;
- **T – Time**, CHECKED, noted, ETAs revised, to/from waypoint, timer set;
- **F – Fuel** – CHECKED, correct tank (selector on both) remaining flight time/time to diversion considered;
- **I – Instruments** – SET AND CHECKED, suction, amps, annunciators;
    - **Icing** – CONSIDERED, carb ice/engine ice as required
- **R – Radios** – SET AND CHECKED, required main and standby frequencies set;
- **E – Engine** – CHECKED, temperatures and pressures, mixture richen, cowl flaps closed/as required.

Note: HATFIRE may also be used as an enroute check the same way as SAFDIE is described in the Cruise section, as the 'FIRE' part encompasses the same items, only in a different way.

Approach (or downwind) checks can be completed as checks or in a flow pattern followed by a check-list, as available.

↣ Typical approach/downwind checks are as follows **(BUMPFFEL)**:
- **Brakes** – ON check pressure and ensure OFF;
- **Undercarriage** – FIXED;
- **Mixture / Pitch / Power** – SET;
- **Flaps** – as required;
- **Fuel Valve** – CORRECT TANK (BOTH or LEFT/RIGHT - as installed), CHECKED;
- **Engine's Temperature & Pressure** – CHECKED;
- **Landing Light** – ON.

Approach and Landing

When arriving at an airfield in anticipation of an approach, aim to be configured at a speed, height and power setting which will allow an easy transition from cruise to approach. A recommended configuration for a normal arrival would be at circuit altitude, with approximately 90-110kts and around 18-20" manifold pressure. The speed should be 5-10kts below the speed for the selection of the first stage of flap. A lower speed will also increase safety margins if you are entering a busy circuit.

A normal approach for landing should be flown at a speed of approximately 75-85kts (90-100mph) flap up, reducing to 65-75kts (75-90mph) with full flap. It is important to ensure the approach is planned and executed to ensure a stable descent profile at the correct speed is achieved.

Approaching with higher than recommended speed will lead to a prolonged flare and/or several bounces down the runway if the aircraft is forced to land too early. A lower speed may lead to loss of elevator control, especially on early models, which have a smaller elevator, and at a forward Centre of Gravity, or a hard landing from an excessive rate of descent on touchdown.

To prevent landing errors resulting from the approach angle or speed, the aircraft should be flown in a manner that ensures a stable approach, with a speed of approximately $1.3Vs$ + wind correction is maintained from approximately 500 ft to 50ft over the runway elevation.

Once established on final, in the landing configuration, final approach checks must be completed. These comprise vital actions that must be completed before landing or go-around.

Generally final approach checks in a single pilot VFR operation should be completed from memory, however a control column checklist is a suitable alternative.

Typical final approach checks are as follows (CUMP):
- **Cowl Flaps** – OPEN;
- **Undercarriage** – FIXED;
- **Mixture** – SET for go-round;
- **Pitch** - FULL FINE.

Final Approach Speed

The recommended minimum approach speed of 65kts with 40 degrees of flap, is only provided at maximum weight in most POHs. When flying a very light aircraft, it should be remembered that there will be a tendency to float.

Minimum approach speeds where performance allows, can be increased by approximately 5kts to provide a bigger safety margin. In windy/gusty conditions, a wind correction factor should also be applied increasing the safety margin again to allow for wind shear.

The rule of thumb for application of a gust factor is to add ½ the Head Wind Component (HWC) and all of the Gust to the minimum approach speed, Vref.

Eg: A reported wind of 060/20G30 on Runway 36 provides a HWC of 10kts, and a Gust factor of 10kts, therefore the addition should be 15kts. Although this sounds like a large increase in speed the following must be remembered, only head wind component must be considered and as only half is taken, with a steady wind there is still a reduction in distance from the reduced ground speed.

When the wind is gusting there is normally always a headwind factor, which would not be taken into account for performance planning, so even where there are strong gusts, landing performance should not be significantly affected (e.g. in the above example the reference ground speed is only increased by 5kts). Additionally performance factors should be applied to ensure there is adequate margin for error and minimum safety levels in calculation to permit variations in conditions. It is important to remember gusty conditions can also lead to wind shear, which can significantly affect touch down speeds, and therefore landing distance, if there is a sudden loss of headwind, or sudden tail wind component on landing.

Whenever the wind is reported gusting, windshear and turbulence is almost always present and it is advisable to have some extra speed to cope with these conditions. The rule, however, is only a starting point, and must be modified as required for conditions and field length. The maximum speed increment shall be not higher than 15 knots to be able to use the landing distances provided by the Cessna in the Performance Section of the POH.

Short Field Landing

For a short field operation the exact speed is specified in the flight manual for the weight and conditions applicable. Positive control of the approach speed and descent should be made to ensure accuracy of the touchdown point.
The landing should be positive, with a high nose attitude (on the main wheels) and as close as possible to the stall.

Crosswind Landing

When approaching to land with a crosswind the aircraft flight manual discusses crabbed, slipping or combination method.
To prevent drift on finals the aircraft should be crabbed into wind as detailed above.
For landing, the aircraft nose should be brought in line with the runway. In doing so the aircraft will begin to drift, and the 'into wind' wing will need to be lowered just enough to keep the aircraft on the runway centre line. The 'into wind' wheel will therefore make contact first, thereafter the remaining main wheel and then the nose wheel should be positively placed on the ground, and ailerons placed into wind to prevent aerodynamic side forces.
The question of differing techniques is therefore only a question of where to transition from the 'crabbed' approach to the landing configuration.
This is ideally achieved in the round out, however it may be commenced earlier to assist the student with learning the degree of, control input, to apply.

In a strong crosswind a slightly higher approach speed may be required to maintain more effective control against the wind factor. A slightly higher touchdown speed is also recommended to prevent drift in the transition between effective aerodynamic control and effective nose wheel steering (runway length permitting).
Reduction in flap setting improves lateral stability for added crosswind control should the student meet conditions he/she feels beyond his/her capabilities.

It should be noted the C206 is controllable with full flap in excess of the maximum demonstrated crosswind, and is a good exercise to practice with an instructor.

The C206 main landing gear wheels are positioned just in front of the rear passengers seats leaving the aircraft with reasonably poor directional control. When landing in crosswind conditions, especially with an aft centre of gravity, touchdown must be preformed with aircraft nose perfectly aligned with the runway centreline. Weather cocking and poor nose wheel control is exaggerated in such conditions and any misalignment will cause the aircraft's tail to move forward, and the aircraft will veer dangerously away from the centreline. This may lead to pilot induced directional oscillation, possibly leaving the aircraft resting in the bushes next to the runway – a very undesirable situation!

Flapless Landing

Two items of importance should be considered for a flapless landing.
1. Lack of drag to assist with the descent and approach.
2. The increased stall speed compared to the normal landing configuration.

To assist with overcoming these items a slightly lower power setting and higher approach speed should be used.
For power management, it is recommended to extend downwind slightly to prevent the need to use a very low power setting whenever possible. The approach slope will be shallower than normal, with a higher nose angle.
The round out will be flatter than for a normal approach, because of the attitude, and the higher speed and lower drag combination will tend to cause the aircraft to float.

Care should be taken with assessing performance and required landing distances without flap in the case of a need to land on a short runway. Remember that technique can play a significant role in the resulting performance. For a short field landing, the increase in final approach speed should be by the margin between the stall with full flap and with flap up, since the short field approach speed is based on 1.3x the stall speed in the approach configuration.

Balked Landing

The wing flaps should be reduced to 20 degrees immediately after full power is applied, and an airspeed of 80kts (90mph) maintained.

If an obstacle must be cleared during the go-around and associated climb, the wing flaps shall be left at 20 degrees and a safe climb speed (Vx) maintained until the obstacle is cleared. Above 3000 feet altitude, the mixture should be leaned in accordance with the fuel flow placard to obtain maximum power.

Upon reaching a safe altitude, the flaps should be retracted in stages to the full UP position, and the cowl flaps confirmed open.

After Landing Checks

When clearing the runway after landing, it is vital to complete the after landing checks for engine management and airmanship considerations.

For engine handling considerations, Checks must include the cowl flaps (if applicable) and at higher altitudes or temperatures, the mixture which has been set rich for the go-around, should be leaned for taxi to prevent spark plug fowling.

The wing flaps must be retracted (to prevent ATC suspecting a hijacking has occurred!), it is polite to select the strobe and landing lights off and the transponder should be selected to standby, unless otherwise dictated by ATC procedures.

After Landing checks can be completed in a flow pattern followed by a check-list, or as a check list where available.
Typical after landing checks are as follows:
- **Cowl Flaps** – OPEN for taxi;
- **Mixture** – SET for taxi;
- **Flaps** – UP;
- **Strobes and Landing Light** – OFF;
- **Transponder** – STANDBY.

Taxi and Shutdown

Taxi should be planned to suit engine cooling requirements when needed. If you are operating on rough gravel remember to avoid needing to operate the aircraft stationary at idle for prolonged periods.
In a normally aspirated engine, providing the approach was accomplished without using excessive amounts of power, in most cases the taxi should provide sufficient time for cooling down the engine. For a turbo additional cooling may be required (see more in the following section on Engine Handling Tips).

Before completing the shutdown, it may be desired to complete a dead-cut check to ensure all magneto positions, in particular the OFF position is working, so the propeller is not left 'live'.

Shutdown again can then be accomplished in a flow pattern, followed by a checklist where available, to confirm all items are complete prior to vacating the aircraft.

✈ Typical shutdown checks are as follows:
- **Avionics** – OFF;
- **Mixture** – CUTOFF;
- **Magnetos** – OFF;
- **Master** – OFF;
- **Control Lock** – IN;
- **Flight Time/Hour Metre** – RECORDED;
- **Tie Downs/Screens/Covers** – FITTED.

Circuit Pattern

The circuit pattern may differ from airport to airport. Ask your instructor, the briefing office, or consult the relevant aeronautical information publication for information on the pattern on your airfield.

The standard circuit pattern, unless published otherwise, is the left circuit pattern at 1000ft above ground for piston engine aeroplanes.

The circuit pattern contains all the critical manoeuvres required for a normal flight, condensed into a short space of time. It is a great way to learn the critical flight checks, practice manoeuvres and improve overall flying skills.

The following provides guidelines for circuit operations, *(Note: a summary of all applicable checks from previous paragraphs is included to permit use of this section as a concise study guide):*

➜   Complete the aircraft preflight walkaround, ensuring fuel and oil quantities are sufficient, all required equipment is serviceable, and the condition of the aircraft and all components is acceptable for flight, and complete the before start checks:
- **Preflight Inspection** – COMPLETE;
- **Tach/Hobbs/Time** – RECORDED;
- **Passenger Briefing** – COMPLETE;
- **Brakes** – SET/HOLD;
- **Doors** – CLOSED;
- **Seats / Seatbelts** – ADJUSTED, LOCKED;
- **Fuel Selector Valve** – BOTH/CORRECT TANK;
- **Cowl Flaps** – OPEN;
- **Pitch** – FULL FINE;
- **Magnetos** – BOTH;
- **Avionics** – OFF;
- **Electrical Equipment** – OFF;
- **Rotating Beacon** – ON.

➜   Once ready to start, with the master ON, complete the 'ready for start' or 'cleared for start' checks:
- **Annunciators** – CHECK (if applicable);
- **Circuit Breakers** – CHECK IN;
- **Mixture** – RICH / AS REQUIRED*;
- **Prime** – AS REQUIRED (50-80lbs);
- **Throttle** – SET approx ½ centimetre**;
- **Propeller Area** – CLEAR.

➜   Start the engine, including completion of the after start flows, and complete the after start and pre-taxi checks:
- **Mixture** – SET for taxi;
- **Magnetos** – CHECKED;

- **Engine Instruments** – CHECKED;
- **Flaps** – RETRACTED/SET;
- **Transponder** – STANDBY/GROUND.

↳ Test the brakes as soon as possible after the aircraft begins moving, then at any convenient time during the taxi check the flight and navigation instruments, then complete the taxi checklist;
- **Brakes** – Checked;
- **Avionics and Flight Instruments** – Checked/Set;
- **Nav Instruments** – Tested/Checked/Set.

↳ Taxi towards the runway and stop the aircraft clear of the runway and in a suitable position to carry out the **Engine Run-up**. Ensure that:
- The slipstream will not affect other aircraft;
- A brake failure will not cause you to run into other aircraft or obstacles;
- Loose stones will not damage the propeller.

↳ Prior to the **Engine Run-up** it is important to check the following items:
- Confirm fuel is on correct tank (always run up on the tank you intend to takeoff;
- Check the mixture is set correctly for the run-up;
- Check temperatures and pressures in the green range.

↳ Set the park brake and complete the **Engine Run-up:**
- **Power** – Set 1700rpm;
- **Mixture** – Set for elevation (above 3000ft density altitude);
- **Magnetos** – Check left, both, right, both, confirm smooth operation within limits for drop and differences;
- **Pitch** – (if applicable) Cycle three times for a cold engine, minimum once if the engine has bee running.
- **Engine's Temperature & Pressure** – Check;
- **DI** – Aligned with compass;
- **Power** – reduce to idle, confirm steady at 500-700rpm, return to 1000rpm.

↳ Complete the **Pre Takeoff Vital Actions** (Too Many 'Pilots Go Fly In Heaven Early):
- **Trims and Controls** – TESTED and SET for takeoff;
- **Mixture** – SET for takeoff;
- **Magnetos** – BOTH;
- **Pitch** – FULL FINE;
- **Gills** *(Cowls)* – OPEN;
- **Gyros** – SET, uncaged (if applicable);
- **Fuel** – SET and CHECKED, contents checked on correct tank, primer locked, pump as required (normally off for Cessna high wing aircraft);
- **Flaps** – SET for takeoff;

- **Instruments** – CHECKED, panel scan check from right to left, DI aligned with compass, check navigation aids for departure, annunciators (if applicable) check clock, note time;
- **Hatches and harnesses** – SECURE;
- **Electrics** – CHECKED, circuit breakers checked, systems set.

↳ Once the before takeoff checklist has been completed, taxi to the holding point and give the required lining up and/or departing calls.

↳ Line up and ensure that the nose wheel is straight (make full use of the runway length available) and perform the **Line-Up Checks (REmember What To Do Last):**
- **Runway** – UNOBSTRUCTED, CORRECT, and nosewheel aligned;
- **Engine** – CHECKED, temperatures and pressures green, mixture set;
- **Windsock** – CHECKED wind direction and strength, (confirm against ATC wind), position control column accordingly;
- **Transponder** – ALT/TA-RA/ON;
- **DI** – ALIGNED with compass and reading correct bearing;
- **Lights** – Landing light and strobes ON.

↳ Takeoff and climb maintaining runway alignment. Keep straight with rudder (will require right rudder due to the slipstream and torque effects);

↳ Hold elevator back pressure to protect the nose wheel, keeping the weight off it, and to allow the correct attitude for flight. Raise nosewheel approximately 50kts in normal takeoff;

↳ Lift off and accelerate to maintain a minimum of 70kts (80mph) for a normal takeoff or 65kts (75mph) for a short field takeoff with 20 degrees of flap;

↳ Upon reaching a safe altitude (300' above airfield elevation), accelerate to a minimum of 80kts (95mph) and raise the flaps (if used);

↳ Complete the **After Take-Off Checks (BUPPMFFEL)**:
- **Brakes** – ON and OFF;
- **Undercarriage** – FIXED;
- **Power / Pitch / Mixture** – SET;
- **Wing Flaps** – UP;
- **Fuel Valve** – CORRECT TANK, CHECKED;
- **Engine Temperature & Pressure** – CHECKEd;
- **Landing Light** – OFF.

↳ At a minimum of 500ft AGL scan the area into which you will be turning and then turn onto crosswind leg using a normal climbing turn (maximum bank angle of 15 degrees or Rate 1);

CESSNA 206 TRAINING MANUAL

Reaching circuit height, level-off, reduce power to approx 2400rpm and 20" manifold to prevent speed increasing beyond the flap limits, select cowl flaps to half or as required for ambient conditions;
- Trim the aeroplane for straight-and-level flight;
- At approximately 45 degrees, and approximately 2nm-2.5nm from the runway, scan the area into which you will be turning and turn onto downwind leg, select a reference point to track parallel to the runway;
- Confirm speed is within the flap limit, and select the first 10 degrees of flap;
- Re-trim for level flight;
- When abeam the centre of the runway, make the downwind radion call and perform **Downwind Checks (BUMPFFEL)**:
    - **Brakes** – ON check pressure and ensure OFF;
    - **Undercarriage** – FIXED;
    - **Mixture / Pitch / Power** – SET;
    - **Fuel Valve** – CORRECT TANK, CHECKED;
    - **Engine Temperature & Pressure** – CHECKED;
    - **Landing Light** – ON.
- Approaching late down wind reduce power to approx 15-17 inches manifold pressure depending on speed,
- After scanning for traffic on base leg and final approach, select a reference point on the wing tip and turn onto base leg using a standard 30 degree bank medium turn;
- Check that the speed is within the flap limit, and lower flap to 20 degrees;.
- Reduce power to 12-15 inches manifold pressure (while keeping the nose up to permit reduction to approach speed), and trim for the descent, speed should not be more than approximately 10kts above desired final approach speed;
- Once the aircraft is trimmed to maintain the desired approach speed, approximately 75-85kts (90-100mph), power should be adjusted to maintain the desired approach angle;
- Visually check the final approach is clear of traffic, and at not less than 500ft AGL, anticipate the turn to final so as to roll out with the aircraft aligned with the direction of the landing runway, (remember radius of turn is dependant on your base speed and drift);
- Once on final select the flaps to landing position, normally full flap, 40 degrees is used, less may be selected if desired for loading or conditions, and make the 'final' radio call;
- Re-trim the aircraft for selected final approach speed, 65-75kts (75-90mph) and the required glide path;
- Complete **before landing checks (CUMP)**:
    - **Cowl Flaps** – OPEN;
    - **Undercarriage** – FIXED;
    - **Mixture** – SET for go-round;
    - **Pitch** – FULL FINE.

→ Execute the appropriate landing procedure;
→ Maintain the centre line during the landing run by using rudder and wings kept level with aileron. Brakes may be used once the nose-wheel is on the ground;
→ Once clear of the runway, stop the aeroplane, set 1000rpm and complete **After Landing Checks**:
  • **Cowl Flaps** – confirm OPEN;
  • **Mixture** – LEAN for taxi (as required);
  • **Wing Flaps** – UP;
  • **Strobes and Landing Light** – OFF;
  • **Transponder** – STANDBY.
→ Obtain the relevant clearance, or make the required radio call and taxi to the parking area;
→ Shutdown and secure the aircraft and complete the shutdown checks:
  • **Avionics** – OFF;
  • **Mixture** – CUTOFF;
  • **Magnetos** – OFF;
  • **Master** – OFF;
  • **Control Lock** – IN;
  • **Flight Time/Hour Metre** – RECORDED;
  • **Tie Downs/Screens/Covers** – FITTED.

Note on Checks and Checklists

Present standard and recommended operating practices on a single-pilot aeroplane dictate use of a checklist AFTER completion of vital actions in a flow pattern on each critical stage of the flight, such as before and after takeoff, on downwind and final leg.

The acronyms above therefore provide a memory aid to allow for completion of the checks prior to reading the checklist. For single pilot operations on light aircraft, acronyms are strongly recommended for memory items and flows. Any convenient acronym is acceptable providing the minimum required items are catered for.
Unless you only ever intend flying one type, it is also recommended to use generic memory items and flows. This will avoid potential omissions when flying different types. A checklist, however, should be specific to an aircraft type and serial number, including any modifications that affect operation.

When a checklist is completed in single pilot operations and no autopilot is available, the checklist should be as hands-free as possible, especially for critical phases. Control column checklists, or a chart clip on the control yoke, are considered the easiest method to achieve this.

The above checks and procedures are based on standard training practices. Application of these checks and development of a checklist for operational use, must be cross referenced against the POH of the aircraft you are flying, and the applicable regulations for the specific operation.

Examples of checklists in document format free for download and editing can be found at http://www.redskyventures.org.

Do-Lists

A 'do-list' or 'read-and-do' list is a type of checklist where actions are completed as they are read. A do-list omits the redundancy built in to a normal check-list procedure since items are only done once, not done then checked.
This type of procedure is often used in training operations and light aircraft for completion of normal checks, however, once a student has reached the solo stage, do-lists should ideally only be used for emergencies/abnormalities and non-standard operations. A do-list is important for non-normal operations, since the procedures are seldom carried out and are too unfamiliar for completion from memory.

Checks for certain emergencies, where time is critical, must be memorised. These are normally followed up by a "do-list" accomplished in a read and do manner. In the later model Cessna POHs and the Cessna quick-reference-handbook which is provided with post 1996 models, memory items are written in **bold** typeface, the remaining abnormal and emergency checks are intended to be completed from the checklist – where possible. See next section for more details on emergencies and abnormal operations.

Flight Operating Tips

The Cessna 206 handles exceptionally well and, unlike her predecessor the C210, suffers from very few handling complexities. Following proper training, the C206 will lead to few problems for the proficient, well trained pilot, providing it is flown according to the POH and accepted standard practices.

The biggest problem that pilots may have with the C206 is overestimating performance. Although the short field and loading capacities are excellent, there are limitations – and pilots who have failed to adhere to the operating performances laid out in the POH have often learned the sad consequences.

The C206, under certain conditions, may have a marginal after takeoff climb, especially with 20 degrees of flap required for a short field takeoff. Although the take-off roll is fairly short, the time and distance required to reach obstacle clearance altitude can be significant, normally total takeoff distances are double the takeoff roll. As with takeoff performance, after takeoff climb performance is worst at high loading, and high density altitudes.

Loading

An all up weight landing, or a landing with a forward centre of gravity, sometimes results in the pilot reaching full aft travel of the elevator before reaching the landing attitude and speed. This may result in a hard landing. This can be particularly hazardous in a short field landing situation, where carrying extra speed on approach for the round out is not possible. Typically this is avoided by ensuring the loading is closer to the aft than the forward limit, which is easily achieved when 4 or more people are on board. Insufficient aft elevator would normally only be a problem when operating with one or two passengers, where any weight should be loaded in the rear, and a weight and balance calculation or load sheet should be completed.

Whenever six adults are carried, the opposite danger appears, in exceeding the aft centre of gravity limit. In this situation, if there is a cargo pod, load the cargo pod first with any heavy items, leaving only hand luggage in the aft cabin compartment. Load smaller passengers in the rear seat, and complete a weight and balance calculation or load sheet. If the aircraft is within limits, but still appears tail heavy, try to move the adjustable passenger seats forward as far as possible, even one notch will make a difference.

Systems Management

If you are upgrading from a less complex aircraft, the additional pitch control, mixture requirements, cowl flaps and fuel pump may be initially daunting. The throttle quadrant concept is useful in dealing with the first three items (see full explanation in the Engine Controls section). It is essential in this regard to establish proper disciplined procedures from the start. If any problems are experienced, the helpful and inexpensive exercise of 'dry flying', running through the procedures in your mind, preferably while sitting in the aircraft is strongly recommended.

The fuel pump should not be confused with operation on some low wing aircraft, and other Cessna types. It is important to remember, the fuel pump should *never* be used in *normal* flight. It is used for starting, fuel surges or vapour locks, and engine driven pump failure.

Engine Handling

The larger engine on the Cessna 206 means mishandling can more easily lead to costly cylinder replacement or untimely failures. It is therefore of utmost importance to ensure proper engine handling practices are adhered

to. Smooth and gradual changes of power, proper warming up and cooling down, monitoring temperatures and pressures, and correct cowl and mixture settings are the most important concepts of correct engine management. Regardless of technique, engine handling must be carried out with these points and their effect in mind to ensure mishandling is avoided.
Cracked cylinders, the most expensive and operationally inconvenient result of poor engine management, are generally caused by rapid heating or cooling, often termed 'shock-heating' and 'shock cooling'. Both are caused by the same phenomena – inadequate time allowed for changes in temperature.

More is often spoken about shock cooling, mainly because in flight the approximately 150kts of cooling airflow over the cylinders means that careless reductions in power can much more rapidly result in sudden temperature losses than increases in power, and also because pilots more naturally think of taking care during increases in power, but forget (of) the serious damage that can occur by uneven contraction of the metals in the engine during temperature reduction.

The following provides some operating guidelines which will help ensure you minimise the results of poor engine temperature management.

Application of Power

Ensure temperatures have risen to acceptable levels to permit full power application, and do not take-off with oil or cylinder temperatures below the green arc (see note below about extreme cold weather).
Apply power slowly and smoothly, take approximately 5 seconds to reach full throttle, whilst monitoring temperatures and pressures.
If conditions permit, stabilise the engine at an intermediate setting, preferably while holding the aircraft against the brakes, approximately 20 inches for normally aspirated and a little more for a turbo (see below), before continuing to takeoff power.

Changes of Power

Always make gradual power changes, planning how much power you require for the desired effect. For example if you want to reduce the speed by 5kts on approach reduce the power by 2-3 inches depending on weight and wait for the desired effect, rather than taking 5 inches off then putting back 3 when you get to the right speed. This method requires 'anticipation' of power changes, and with a little practice will improve your situational awareness with regard to speed and profile on approach.

If a large power change is required, the power change should be stepped, that is reduce the power in more than one stage. When reducing from 23" to 15", for example, first reduce to 20 inches for a few minutes, then 18 inches for a few minutes, to allow cooling before reducing to 15 inches. Depending on the power change and the time permitted by the phase of flight a minimum of one intermediate power setting should be selected.

Note: For situations of windshear, terrain or stall avoidance safety will take precedence over engine handling, and expeditious power application is required.

Power During Descents

A descent often combines increased speed, with reduced power and enriching of the mixture. All three contributing to engine cooling and therefore very important for engine management (see more above in the section on Descent Planning).

Divide the distance, time, or height you have to make the power change by the inches of power reduction plus two. The last two ensures you start power reduction before the beginning of the descent and finish before the end. Begin the first reduction *after* stabilising in the descent, to avoid combining the increase in speed with the power reduction. This can be done by first stabilising the initial power reduction and then starting the descent but this will result in a lower descent speed, depending on how time orientated your flight is. The last power reduction occurs before the end of the descent to allow for speed stabilisation before reaching the circuit, or before reaching whichever phase of flight and power setting your descent from cruise is planned to end.

Mixture Changes

Over enriching of the mixture during the descent will provide extra cooling to the cylinders through the extra fuel introduced. Fear of a lean cut often leads pilots to err on the rich side during descents. A good rule of thumb is one turn per 1000ft, preferably at intermediate times to when the power reduction is made. When an EGT gauge is installed, this method should be carried out in conjunction with monitoring the EGT to maintain it slightly below the cruise setting. The idea is to enrich the mixture in such a way that during descent the resultant mixture is kept the same or slightly below what it was during cruise. Thus gradual reducing of power (Manifold Pressure) during descent will lead to the gradual enrichment of mixture.

Sometimes a procedure of selecting full rich mixture for the missed approach during downwind is taught. At this point the engine may still be developing close to cruise power, and full rich mixture will cause a sudden and unnecessary engine cooling. A better method is to select the mixture fully rich (or to go round setting) on short final when the power in minimum.

Use of Cowl Flaps

Cowl flaps should be used with the above engine management aims in mind, that is maximum time for engine temperature changes, and also to avoid excessively high temperatures. Avoid the temptation to open or close cowl flaps for expediting the engine temperature change desired for this reason.

Fuel and Engine Monitoring

All engines are slightly different when it comes to fuel consumptions. It is recommended that owners and operators institute a fuel monitoring program which enables determination of consumption between refuelling, a process which is now mandatory in most countries for commercial operators. This can be as simple as a requirement to record dipped fuel and delivered fuel quantities in the aircraft log before and after refuelling. To ensure all pilots comply, this is best accompanied by a note stating the requirement in a prominent place. A short review of the previous few flights will enable pilots to confirm performance. Owners or operators may take this a step further by entering the data in a spreadsheet or database. To be of most use the applied cruise power settings should also be recorded.

In addition to fuel monitoring, engine monitoring can also be instituted in the same way. This concept is essential with turbine engines, which are more sensitive to small defects and wear, but also may be extremely useful to piston engines, providing an early warning of engine trouble. Engine trend monitoring is achieved by recording engine temperatures, oil pressure and speed along with altitude, temperature and power settings when the aircraft is stabilised in the cruise.

Extreme Hot and Extreme Cold Weather Operations

In extreme cold weather use of preheat is recommended to avoid engine wear and ensure oil is sufficiently liquid prior to turning over the engine. The details of operation of pre-heat will be contained in the Supplements section for the applicable fitting.

A warm up period of 2-5 minutes should be allowed, thereafter Cessna permits takeoff without an oil temperature indication. It is recommended wherever possible this should be avoided. As a minimum the CHT temperature should be indicating in the green.

To avoid extreme cooling during approach, the profile must be managed correctly to prevent prolonged operation at low power settings.

In extreme hot weather care should be taken to avoid overheating on the ground. Avoid prolonged idling, and high power run-ups. Temperatures, particularly on turbo models, often increase after landing, and in these cases, a low powered approach should be planned to avoid additional time for turbo cooling prior to shutdown (see more below on turbo operations).

In hot weather the fuel system is probe to vapour locks, which can make it difficult to start, and cause fuel starvation during flight. Both require use of the auxiliary fuel pump to purge the fuel vapours. For in flight fuel vapour encounters, additionally a change of fuel tank may be required (see more in the relevant sections on starting, fuel system, and non-normal flight procedures).

Turbocharged Engine Handling

Additional to the above engine handling considerations, a turbo engine needs some extra care.

Over-boosting

On a normally aspirated engine, power can only be increased to just below ambient pressure, hence there is no red line indicated on the manifold pressure gauge. On a turbocharged engine, additional air pressure is supplied from the turbo, so it is capable of producing power well in excess of the maximum permitted for the engine.

A 'waste-gate' is installed in the turbo ducting, which should open when power approaches red-line to prevent the turbo supplying too much power to the engine. The waste-gates, particularly those in Cessna engines, can often fail or are set incorrectly. When a waste gate is not working properly, red-line can be reached at an intermediate throttle setting, and so takeoff power may not require full throttle. Application of full throttle will result in exceeding the engine limits, termed 'over-boosting', causing excessive pressure to the engine.

The only way to prevent over-boosting is to monitor the power application as the manifold approaches the maximum limit, and never apply full throttle until you are sure the limit will not be exceeded.

Spool Up

During initial power application the engine may surge when the turbo 'kicks in'. This is because there is a small delay while the turbo impeller accelerates with the engine speed, resulting in a lag before the throttle setting results in an increase in manifold pressure. When the pressure is available the turbo tries to supply too much to make up for the lag resulting in an engine surge.

The effect of the increasing pressure in the turbo is only felt as the engine approaches ambient, where the pressure begins to be boosted by additional pressure from the turbo. The more quickly this phase is transitioned the more severe the resulting power surge, which in extreme cases may exceed the maximum allowable power.

The best way to avoid damaging surges is to apply power slowly and where possible (runway length or surface permitting) stabilise power just below ambient pressure (20-25 inches depending on altitude) before increasing to takeoff setting.

Cooling Prior to Shutdown

The turbo system runs at extremely high temperatures, often much higher than the engine operating temperatures. If the engine is shut down too early, the resulting shock cooling of the very hot parts in the turbo and the effect of the lack of cooling/lubricating oil to the turbo bearings can cause significant damage and or failure.

To prevent damage caused from early shut-down, the engine should be run at idle power for some time after landing. The longer the engine can be run at idle before shut down the better, but generally 3 minutes is an accepted time. The three minutes may be timed from touch down, so that a long taxi at close to idle can be used as part of the cooling process.

When operating on dirt surfaces, which are damaging to the propeller, and can cause excessive heating in hot weather operations. A low power approach may be planned, then timing can be started from when the power was reduced below ambient pressure during the approach.

Never shut down a turbo engine less than one minute after reducing to idle power unless an emergency condition exists (eg. fire or smoke).

NON NORMAL FLIGHT PROCEDURES

Stalling and Spinning

There is no pronounced aerodynamic stall warning (buffet) on the C206 and an audible stall warning is installed, providing a steady audible signal 10 kts before the actual stall is reached and remains on until the flight attitude is changed. The stall characteristics are conventional for flaps retracted and extended. Slight elevator buffeting may occur just before the stall with flaps down.

Out of balance forces may lead to a pronounced wing drop at the stall.

Spinning is prohibited. If an inadvertent spin is entered recovery is standard.

Spin recovery:
1. Confirm throttle closed;
2. Confirm direction of spin;
3. Simultaneously apply opposite rudder to break spin and forward elevator input to break stall;
4. Once spinning has stopped neutralize rudder and ease out of the dive;
5. To minimize height loss once airspeed is decreasing apply full power and establish in a climb attitude;
6. Accelerate to Vx or Vy as required to regain height loss.

Electrical Malfunctions

An electrical malfunction may be indicated by either the ammeter or over-voltage light.
Electrical malfunctions may be caused by the alternator, voltage regulator, or wiring fault. Whichever the reason, the appropriate abnormal procedure should be followed.
An electrical malfunction will usually fall into one of two categories, an excessive rate of charge or an insufficient rate of charge.

Excessive Rate of Charge

The power supplied to the battery is too high. This may have occurred after excessive electrical demand (e.g. long starting sequence or long taxi at low power with high electrical demand). If a high rate of charge continues for a significant period the battery will overheat and damage may occur to components from the excessive voltage.

The over voltage sensor will automatically shut down the alternator to prevent further damage when charge rates reach 30-31 Volts.

If it is suspected the over-volt situation was temporary, recycle both sides of the alternator switch (OFF then back ON). If normal operation is resumed, continue to monitor the situation and resume flight as normal.
If operation cannot be resumed, or if the fault occurs again, select the alternator OFF to prevent damage from the over-volt condition and continue with the insufficient rate of charge procedure.

Insufficient Rate Of Charge

The alternator is not supplying sufficient charge to the battery:
➢ Switch off the alternator – to prevent further drain to the battery;
➢ Reduce Electrical Load as far as possible;

If the fault finding procedure in the POH is unsuccessful:
➢ Plan to land at the nearest suitable airfield;
➢ If battery power remaining indicates possible loss of all electrical power, inform ATC ahead of time, and make provisions for a flap up landing.

Abnormal Oil Pressure and Temperature

Low oil pressure, which is not accompanied by high oil temperature, may indicate a failure of the gauge or the relief valve or can possibly be a fault with the oil pressure gauge.
This is not necessarily cause for an immediate precautionary landing, but landing at the nearest suitable airfield should be made for inspection of the cause of the problem. The situation should be closely monitored for any changes.

Loss of oil pressure, accompanied by a rise in oil temperature is good reason to suspect an engine failure is imminent. Select a suitable field for a precautionary landing. Reduce engine power as far as possible and plan to use minimum power for the approach. Keep in mind the considerations for continuation in the event of a complete engine failure.

A high oil temperature will result in a small reduction of oil pressure because of the reduced viscosity of the oil. If the reason for the high oil temperature is clear, then the cause should be addressed (for example high ambient temperatures or extended climb). Normal cooling methods, for example reducing the power, richening the mixture, increasing the speed, and ensuring the cowl flaps are open should be used. This situation will require monitoring but, providing the temperature begins to stablise or reduce once the cause is removed, this should not indicate any further problem.

An increase in oil temperature without cause may be a sign of an engine fault and/or impending failure. As with oil pressure faults, a precautionary landing should be planned, while keeping the possibility of a complete engine failure in mind.

Rough Running Engine

A rough engine running can be caused by a number of reasons, faults that can be dealt with from the cockpit include spark plug fouling, magneto faults, fuel vaporisation an engine-driven fuel pump failure, and blocked air intake, see the relevant sections regarding these faults.

Magneto Faults

A a small drop in power or engine roughness, is usually an indication of a magneto fault. Switching from BOTH to the L or R position will confirm if one magneto is faulty.
Take care with switching, as if one magneto has grounded or failed completely, no change will occur on the working magneto, however a complete power loss will occur on the failed magneto.
As with spark plug faults, unless extreme roughness is experienced in the BOTH position, it is preferable to continue operating in this position, and land at the nearest suitable airfield.

An excessive differential or smooth but excessive drop during the engine run-up when switching to one magneto is also an indication of a magneto fault and the flight should be discontinued.

Spark Plug Faults

A spark plug fault can be caused by fouling – carbon deposits or other foreign matter closing the spark gap and preventing a spark, or a spark plug failure, that is, although the plug is clean and apparently functional, no spark occurs, due to an internal fault.

Spark Plug Fouling

A slight engine roughness can be caused by one or more spark plugs becoming fouled. This often occurs during prolonged operation at low power settings with the mixture set too rich, and commonly happens at high density altitudes during taxi, well below 3000ft pressure altitude where Cessna recommends leaning the mixture.

If the fault is due to fouling, leaning the mixture to peak or just rich of peak and running at a moderate power setting for a few minutes to burn off the excessive carbon should fix the problem. Noting that it is not recommended to operate at peak for more than 55% power, however there may be cases where more power is needed, care should be taken to monitor the cylinder temperatures.

Switching to one magneto can normally isolate the problem, as running the affected cylinder(s) on the one faulty plug in each cylinder will cause the affected cylinder(s) stop firing. (This is the same procedure as that applied when an excessive magneto drop or rough running is experienced during the engine run-up prior to departure). As with magneto faults, care should be taken when applying this procedure in-flight, as if fouling is severe more than one cylinder, it is possible that there could be an severe loss of power or engine cut when switching to one plug.

Engine roughness or misfiring on one or more cylinders in the both position, can indicate that two plugs in the affected cylinder(s) have fouled.

The primary cause of spark fouling is mixture setting, however in some cases it may be caused by fuel contamination, or improper fuel type or grade.

Spark Plug Failure

If the problem is caused by a faulty plug or a magneto, then leaning will not help. If the problem persists after several minutes operation at the correct mixture setting, it is likely to be a plug or magneto fault. Continue to operate on BOTH, or if extreme roughness dictates selection of the L or R position for smoother operations, remain operating on in this position and continue to the nearest suitable airfield.

Engine Driven Fuel Pump Failure

An engine driven pump failure can be identified by a sudden drop in fuel pressure, *followed by* a loss of power while operating from a fuel tank with adequate fuel supply. (Note – a similar indication will occur with fuel starvation).

Following the power loss, immediately hold the Red, left side of the auxiliary fuel pump in the HI position to re-establish flow. If the power loss occurs on takeoff, continue to hold the fuel pump in the HI position until you have reached a safe altitude and power is reduced.

In cruise, once flow is re-established the normal ON position (the Yellow side of the switch) should be sufficient for continued operation. If power continues to be interrupted reselect and hold the HI side of the switch as required.

Plan to land at the nearest suitable airfield.

Excessive Fuel Vapour

Significant problems have occurred on the C200 series with fuel surges caused by fuel vaporisation, often leading to engine failures and forced landings, however very little is written about it.

The POH recommends for "excessive fuel vapor", a fuel stabilisation procedure, when fuel flow fluctuations of "1Gal/hr or more or power surges" occur. Initial actions require turning on the fuel pump, resetting the mixture and changing tanks.

It should be noted, that the predominant cause of fuel vapour related engine surges or stoppages in pre-1996 models, is from fuel vaporisation in the reservoir tank from the engine driven fuel pump return line (see schematic in the Fuel System section). In this case selecting an alternate tank is critical in resuming normal fuel flow, as selecting only the auxiliary fuel pump may make the problem worse by increasing the flow to the engine driven pump, and so increasing the fuel return to the respective reservoir, which is most likely the cause of the vapours.

Blocked Intake Filter (with Alternate Air Source)

In the event of a blocked intake air filter with a manual alternate air source selector, selection of the alternate air will bypass the fault and resume normal flow of *unfiltered* air to the engine. Land at the nearest airfield with a maintenance facility, where possible, to inspect and correct the cause of the blockage. Take care when operating into a dirt strip on the alternate air source. If the blockage was experienced in icing conditions it is possible that it was caused by a build up of ice on the filter. Once icing conditions no longer exist normal air can be reselected to check if the fault has been cleared.
(An alternate air source should not be confused with the alternate static air, see below).

Inadvertent Icing Encounter

If the aircraft is equipped for icing, activate all installed icing equipment.

If icing persists beyond capabilities of equipment or no equipment is installed, climb or descend to a level where icing is not encountered.

If this is not possible, alter the course to expedite departure from icing conditions. Increase power to aid in preventing significant build up on the propeller.

If icing persists on the airframe during approach, plan a flap up landing to avoid disturbance in airflow adversely effecting elevator control.

Static Source Blocked

A blocked static source will be indicated by a failure or misreading of the Altimeter and VSI and a slight misreading of the air speed indicator.

If an Alternate Static Source is available:
✈ Select Alternate Static Source

If an Alternate Static Source is not installed
✈ IMC – if in an emergency, when weather requires and no alternate means of altitude indication is available, consider breaking glass in one of the static instruments to regain static source.
✈ VMC – inform ATC of failure of altitude information, continue to the nearest airfield.

EMERGENCY FLIGHT PROCEDURES

General

The main consideration in any emergency should be given to flying the aircraft.
Primary attention should be given to altitude, attitude and airspeed control and thereafter to the emergency solution.

Rapid and proper handling of an emergency will be useless if the aircraft is stalled and impacts with the ground due to loss of control. This is most critical during takeoff, approach and landing.

The checklists in this section should be used as a guide only. The emergency checklist and procedures for your particular aircraft model specified in the aircraft's Pilots Operating Handbook should be consulted for operational purposes.

Emergency During Takeoff

Any emergency or abnormality during takeoff calls for the takeoff to be aborted.
The most important thing is to stop the aeroplane safely on the remaining runway.

After the aircraft is airborne, re-landing should be considered only if sufficient runway is available for this purpose. As a general rule, the runway is sufficient, if the end of the runway can be seen in front of the aircraft. Where no sufficient runway is available, the engine failure after takeoff procedure should be followed.

Engine Failure

The recommended gliding speed is 75-85kts (85-100mph), depending on weight. This speed should be used in all situations that require glide performance, unless the manual specifies otherwise.

The majority of fatalities resulting from engine failures are caused by loss of control. It is important to remember, and this cannot be said enough times, that regardless of what else you accomplish it is vital that airspeed is maintained throughout execution of the emergency, up to the point of flare and landing. Even in the most unsuitable terrain you have a reasonable chance of a successful outcome – that is all occupants alive – providing the aircraft is maintained in a controlled state until the landing/ditching is made.

Engine Failure after Takeoff (EFATO)

In the event of an engine failure after takeoff first fly the aircraft:
↦ **Promptly lower the nose and establish a glide attitude to maintain 75kts (90mph).**

Landing should, wherever possible, be planned straight ahead and within ±30° to either side of the flight path.

The turn, if required, should be made with no more than 15° of bank.

If sufficient height exists, Select the Auxiliary Fuel Pump on HIGH and change fuel tank selector where required. A high number of engine failures occur due to fuel mismanagement or fuel starvation.

If adequate time exists, secure the fuel and ignition system prior to touchdown to reduce the possibility of fire after landing, in accordance with the procedure below.

Any attempt to restart the engine depends on altitude available.

A controlled descent and crash landing on an unprepared surface is far more preferable to uncontrolled impact with the ground in the attempted engine start.

EFATO Procedure:
↦ **Airspeed** - 80kts (95mph) flap up
    - 70kts (80kts) flap down
  This speed is the recommended speed for an engine failure or forced landing
↦ **Fuel Selector** – CHANGE;
↦ **Fuel Pump** – ON HIGH;
↦ **Flaps** – FULL.

When sufficient time exists, secure the engine:
↦ **Mixture** – IDLE CUT-OFF
↦ **Fuel Selector** – OFF;
  This will ensure that the engine will be cut-off from the fuel system and thus minimise fire possibility after an impact.
↦ **Ignition Switch**- OFF;
↦ **Master Switch** – OFF
  The master switch should be switched off after the flaps being set in the desired position, to minimize the chance of a fire after touchdown.
↦ **Doors** - UNLOCKED
  The doors should be unlocked in aid of rapid evacuation after the touchdown.

After landing:
- ✈ Stop the aeroplane;
- ✈ Check that fuel, ignition and electronics are OFF;
- ✈ Evacuate as soon as possible.

Gliding and Forced Landing

For a forced landing without engine power a gliding speed of 85kts (100mph) should be used, however if maximum distance is not required a higher speed may be maintained to allow for turbulence and wind penetration.

The first priority is to establish glide speed and turn toward the suitable landing area.

While gliding toward the area, the cause of the failure should be established. An engine restart should be attempted as shown in the checklist below.
If the attempts to restart the engine fail, secure the engine and focus on completing the forced landing (further attempts to restart distract the pilot from performing the forced landing procedure).

If the cause of engine failure is a mechanical failure or fire, the engine should be secured immediately and no restart should be attempted.
If the failure is partial, resulting in reduced or intermittent running, it is recommended to use the partial power till arrival overhead the intended area of landing. Thereafter the engine should be secured when the forced landing procedure is initiated. If a partial power setting is used and power is lost or suddenly regained during the forced landing circuit, this may change the gliding ability of the aircraft so dramatically, that it will be impossible to reach the intended landing area safely.

Forced landing procedure, initial actions:
- ✈ Convert speed to height;
- ✈ Trim for 80kts (90mph)*;
- ✈ Select a field, plan the approach.

*Note the best glide speed is normally 75kts at MAUW, the recommended FLWOP speed is slightly higher presumably for manoeuvrability where maximum distance is not required, and to provide a safety margin for speed reductions.

Finding the fault and restart:
(Note: many good acronyms are available for fault finding checks, the one below uses the MAFIT acronym)

- **Mixture** – FULLY RICH;
  Mixture is recommended to be set rich in the Pilot's Operating handbook, however if it is suspected the cause of engine failure is from too rich setting at altitude, leaning can be opted for.
- **Alternate Air** - (if installed*)– PULL;
- **Fuel** – Selector ON Correct Tank, Contents Sufficient, Primer IN AND LOCKED, Pump – activate 3-5 seconds at ½ throttle then OFF;
- **Ignition** – CHECK L/R and BOTH, Select START if the Propeller is not wind-milling;
- **Throttle** – ADVANCE SLOWLY in attempt to RELIGHT.

*Some models have a manual selector for alternate air in the event the intake becomes blocked, other models have a spring loaded intake providing automatic bypass of the filter by air pressure.

Securing the engine:
- **Mixture** – IDLE CUT-OFF
- **Fuel selector** – OFF;
  This will ensure that the engine will be cut-off from the fuel system and thus minimise fire possibility after impact.
- **Throttle** – FULLY FORWARD;
- By opening the throttle any fuel left in the engine will be sucked out, and the fire possibility will be minimised.
- **Ignition switch**- OFF;
- **Master switch** – OFF;
  The master switch should be switched off after the flaps are selected to the desired position for landing, to minimise the possibility of an electrical fire.
- **Doors** - UNLOCKED.
  The doors should be unlatched in anticipation of a quick evacuation after the touchdown. After landing the same procedure as detailed for an engine failure after takeoff above, should be initiated.

After landing:
- Stop the aeroplane;
- Check that fuel, ignition and electronics are OFF;
- Evacuate as soon as possible.

Simulated Forced Landing
In case of simulated forced landing training, sufficient time should be allowed to cool the engine down before reducing to idle power.

Where a surprise forced landing is desired for instruction or testing, this can be achieved by completing a series of slow flight manoeuvres to avoid arousing student suspicion.

During an extended glide, select partial power for a brief period every 500-1000ft to provide engine warming and to ensure power is available (an engine failure will not be noticeable with a windmilling prop).
Avoid transitioning from an extended glide to full power. If the missed approach is initiated early, selection of partial power can be made and the descent converted into a 'slow safe cruise' configuration. This can be useful in converting the forced landing exercise into a precautionary landing exercise with a low level inspection followed by a simulated short field circuit and overshoot onto the same runway.

Engine Fire

In case of fire on the ground, the engine should be shut down immediately and fire must be controlled as quickly as possible.
In flight such an emergency calls for execution of a forced landing. Do not attempt to restart the engine.
A sideslip may be initiated to keep the flame away from the occupants, this procedure can be also used to extinguish the fire.
If required, the emergency descent may be initiated to land as soon as possible. Opening the window or door may produce a low pressure in the cabin and thus draw the fire into the cockpit. Therefore, all doors and windows should be kept closed till short final, where the door should be open in anticipation of a quick evacuation after the landing.

An engine fire is usually caused by a fuel leak, an electrical short, or an exhaust leak.
If an engine fire occurs, the first step is to shut-off the fuel supply to the engine by putting the mixture to idle cut off and fuel valve to the off position.
The ignition switch should be left on and throttle fully open in order for the engine to use the remaining fuel in the lines.

If an engine fire occurs, the following checklist should be used accurately and expeditiously.

During an engine start on ground:
- ✈ **Cranking** – CONTINUE, in attempt to get a start.
  This will suck the flames and burning fuel into the engine.
  Note: starter limits are of lessor concern to fire damage if the threat still exists.

If the engine starts (and flames extinguish)
- ✈ **Power** – 1700rpm for a few minutes;
- ✈ **Shutdown** – ACCOMPLISH and secure the engine.

If the engine does not start:
- **Throttle** – FULLY OPEN;
- **Cranking** – CONTINUE;
- **Engine** – SECURE:
  - **Mixture** – IDLE CUT OFF;
  - **Fuel SELECTOR** – OFF;
  - **Ignition switch** – OFF;
  - **Master switch** – OFF.
- **Fire Extinguisher** – **APPLY, us**e the fire extinguisher, and wool blankets if the fire persists.

Once Flames are Extinguished and Engine Secure:
- **Engine** – INSPECT for damage.
  Do not restart and call for maintenance for the repairs.

Engine Fire In flight:
- **Mixture** – IDLE CUT-OFF;
- **Fuel valve** – OFF;
- **Throttle** – FULLY OPEN;
- **Master switch** – OFF;
- **Cabin Heat and Air** – OFF (To prevent the fire from being drawn into the cockpit).
- **Airspeed** – 120kts (140mph) If the fire is not extinguished, increase to a glide speed which may extinguish the fire.
- **Forced landing** – EXECUTE.

Electrical Fire

The indication of an electrical fire is usually the distinct odour of burning insulation.

Once an electrical fire is detected, attempt to identify the source. If the affected circuit or equipment cannot be identified and isolated, switch the master switch off, in an attempt to remove all possible sources of the fire.

If the affected circuit or equipment is identified, isolate the circuit by pulling out the applicable circuit breaker and/or switching the equipment off.

Smoke may be removed by opening the windows and the cabin air control. However, if the fire or smoke increases, the windows and cabin air control should be closed.

The fire extinguisher may be used, if required. Ensure the cockpit is adequately ventilated after using the fire extinguisher, to remove the gases.

Landing should be initiated as soon as practical at the first suitable airfield.

If the fire cannot be extinguished, land as soon as possible.

Emergency Exit Procedures – Cargo Version

If it is necessary to use the cargo doors as an emergency exit with the flaps extended, open the doors in accordance with the instructions shown on the placard, as detailed below.

EMERGENCY EXIT OPERATION
1. Rotate forward cargo door handle full forward and then full aft
2. Open forward cargo door as far as possible
3. Rotate red lever in rear cargo door forward
4. Force rear cargo door full open.

🕮 It is of vital importance to ensure this placard is available, positioned correctly, and clearly legible, in accordance with the limitations section of your Pilot's Operating Handbook.

When travelling with passengers, particularly if the flight is over water, and if the passengers are not familiar with the aircraft, be sure to brief on the emergency exit procedure. Although the procedure sounds complicated, it is quite logical in practice and a demonstration in this regard will be most effective.

In the event there is a delay or difficulty in opening the door it may be necessary for all passengers to exit through the front door.

PERFORMANCE SPECIFICATIONS

Performance figures, unless otherwise specified, are given at maximum load (3600lbs) and as an indicated air speed.

Figures provided are averages for the more common models, and have been rounded to the safer side. Performance varies significantly between models, the average or most common figures are indicated. REMEMBER these figures may not correspond to those for your particular model, ALWAYS Confirm performance and operating requirements in the approved aircraft flight manual before flying.

Structural Limitations

Gross weight (take-off and landing)	3600lbs (1966 and later)
Standard empty weight	1850lbs - 2500lbs
Maximum baggage allowance in aft compartment	180lbs
Flight load factor (flaps up)	+3.8g – -1.52g
Flight load factor (flaps down)	+2.0g – 0g

Engine Specifications	Maximum (5 minutes only)	Maximum Continuous
Engine (Continental IO-520 series) power	300BHP at 2850rpm	285BHP at 2700rpm
Engine (Continental TSIO-520 series) power	310BHP at 2700rpm	285BHP at 2600rpm
Engine (Lycoming TSIO-540 series) power	325BHP at 2700 rpm, (flat rated) maximum continuous	
Oil capacity	12Qts normally aspirated engines, 13Qts Turbo and External Filter engines Do not operate on less than 9Qts* minimum	

*Curiously, many models of C210 and C206 have the same engine whilst Cessna recommends a minimum of 7Qts for the C210 and 9Qts for the C206. This may be due to the tendency of C206s to sit tail low, as an inaccurate reading will be obtained if the engine is not relatively level. However, most aircraft maintenance engineers recommend to operate on the high side for the C210, and the low side for the C206.

CESSNA 206 TRAINING MANUAL

Fuel		
Usable fuel early models before 1979	Standard tanks	63USG or 59USG*
	Optional Long range	80USG or 76USG*
*The U206G from 1977-78 had the reduced capacity listed here due to a modified bladder wing tank design which was replaced in 1979 with the integral tanks.		
Useable fuel 206G after 1979 and 206H	Standard Tanks	88USG
	Filler cap quantity	30USG/115lt per side
All models	Optional Tip tanks	Additional 16 USG/60 litres

Landing Gear Pressure	
Main wheel tyre pressure	35psi 8ply, 42 psi 6ply
Nose wheel tyre pressure	29-49 psi depending on type (refer manual)
Nose strut pressure	80 psi

Maximum Speeds		
Never Exceed Speed, (Vne)	183kts (210mph)	(top red line)
Maximum structural cruise speed (Vno)*	149kts, (170mph)	(top of green arc)
Maximum demonstrated crosswind component**		15kts (20mph)
Maximum manoeuvering speed (Va)		120-93kts (mph)
*May not be exceeded unless in smooth air conditions **Late models only		

Flap Limitation Speeds:		
Note: Flap limitation speeds vary slightly with model.		
0-10 degrees	140kts (160mph)	(Placarded on Flap Lever)
10-40 degrees	100kts (115mph)	(top of white arc)

Stall Speeds			
Stall speed, clean (Vs)	55kts	(63mph)	(bottom of green arc)
Stall speed, landing config. (Vso)	46kts	(52mph)	(bottom of white arc)

Performance for Normal Operations

Normal take-off, flaps 0-20°	Raise nose at 50kts (60mph), Accelerate 70-80kts (80-95mph) Retract flaps once obstacles are cleared	
Short field take off, Flaps 10°	Accelerate tail low, airborne maintain 65kts (75mph) until clear of obstacles Accelerate 80kts, (90mph) before retracting flaps, clear of obstacles accelerate to Vy	
Fuel Placard – Takeoff Fuel flow	2850rpm	2700rpm
Sea Level	144lbs/hr	138lbs/hr
4000ft	132lbs/hr	126lbs/hr
8000ft	120lbs/hr	114lbs/hr
Best rate of climb speed (Vy)	Sea level	85kts (100mph)
	10,000ft	79kts (95mph)
Best angle of climb speed (Vx)	75kts (90mph)	
Normal climb out speed	Initial 70-80kts, (80-95mph)	Enroute 95-105kts, (110-120mph) or as required for performance
Normal approach flaps 40°	65-75kts, (75-90mph)	
Normal approach flaps up	75-85kts, (90-100mph)	
Short field landing	64kts, (75mph)	

See more on short field performance and speeds in the Normal Operations section

Speeds for Emergency Operation

Engine failure after take-off	80kts (95mph) flaps up
	70kts (80mph) flaps down*
Best glide	2800lbs 65kts (75mph) flaps up
	3200lbs 70kts (80mph) flaps up
	3600lbs 75kts (90mph) flaps up
Landing without engine power	80kts (95mph) flaps up
	70kts (80mph) flaps down*
Precautionary landing	70kts (80mph) flaps up
	75kts (90mph) flaps down*

*Note: Cessna does not specify which flap settings apply for takeoff and landing emergency speeds in most texts. Rounding the speeds up would allow better safety margins, therefore it is assumed that takeoff would be 20 degrees and landing 40 degrees of flap respectively.

Approximate Cruise Performance Figures

Continental IO520 series 300hp engines, U206G, ISA conditions, 2400rpm (as detailed in the POH bperformance pages)

Cruise at 2000ft pressure altitude	23"MP, 131KTAS, 13.3USG/hr
Cruise at 6000ft pressure altitude	23"MP, 138KTAS, 14USG/hr
Cruise at 10,000ft pressure altitude	20"MP, 135KTAS, 12.4USG/hr

Block Planning Figures*

Block Cruises (non-turbo) *Recommended figures for planning performance estimates*	2400rpm, 23" or available MP
	120kts TAS
	60lt/100lbs/15USG per hour
	76-88USG tanks 4hrs, 59-63USG tanks 3hrs safe endurance + reserve

Cruise figures provided from the pilot's operating handbook should be used with a contingency factor, block cruises speed and fuel flow allow for contingency and for climb and descent, and are normally applied for planning purposes. Where field/operations approach limiting values, performance figures must be consulted to confirm performance.

GROUND PLANNING

For in-depth ground planning, the figures in the flight manual for the aircraft you are flying must be used. For approximate calculations, block figures may be used.

Block figures must provide built in error margins, for example speeds must be lower and fuel consumptions higher that those normally experienced, and to those specified in the POH. These figures should normally allow a margin for error of approximately 10% over the POH figures. Sample block figures are provided above for a normally aspirated C206, but remember, this will *not* apply to all models.

Block figures are a simple method for estimating performance, but should NEVER be used where the performance is critical.

When calculating cruise performance and fuel consumption from the flight manual, remember that the figures indicated are for a new aeroplane, and typically a minimum of 10% contingency (a factor of 1.1) should be applied to all figures where no other safety factor is applied. It is a recommended to apply a minimum 10% contingency to all fuel calculations, and in some countries 10% contingency fuel is required to be carried by law.

It is also important to remember to use the flight manual from the aircraft you are flying when calculating performance. As illustrated clearly in this manual, different models and different modifications on the same model can vary the performance significantly between different aircraft serial numbers.

Most POH's will have graphs similar to the ones included below, later models provide slightly improved versions, however performance graphs can vary significantly, especially if a modification such as an engine or STOL kit has been completed. In this case the graphs will be provided by the approved Supplemental Type Certificate (STC) from the company offering the modification. These graphs should have been duplicated in the performance section, however some operators erroneously leave them in the supplement section at the back of the operating handbook or place them in a separate folder all together.

When calculating performance, ensure all instructions and foot notes are read, as these can have an effect on the graph or table's interpretation and use. Thereafter tables must be read using the appropriate ambient data and weights. For example, do not use airfield elevation when pressure or density altitude is required. In the graphs below, pressure altitude is applied at a range of temperatures, however in some graphs only one temperature is provided for each altitude, therefore it must be deduced that the required altitude figure is density altitude. This could potentially lead to an error of 1000ft per 8 degrees above standard.

The units (pounds versus kilo's or moment versus moment/1000) should never be confused or mixed up. As a reminder, lest we think we are infallible, this error has already resulted in a forced landing in a Boeing 767 in Canada following the conversion to metric.

Where interpolation is required for performance, it is always acceptable to use the next highest figure for fuel, weight, altitude and temperature, or the next lowest rate of climb or cruise speed.

Performance figures for takeoff and landing in the flight manual, are only provided for paved dry runways, with performance degradation figures sometimes provided for dry grass runways. It is important to remember that wet, and sandy or marshy conditions are not allowed for anywhere. Any estimate should be made considerably on the safe side, recommended figures are from 1.1 to 1.3 depending on the degree of additional friction, and may be considerably more if there is a combination of figures. See further on this matter in the tables following.

Lastly, when calculating take-off and landing performance, once all the external variables have been taken into account, remember that pilot performance factors must also be considered. A good rule of thumb to apply is a factor of 1.33 for take-off and 1.43 for landing. These factors are normally required by the relevant national aviation legislation for commercial operations, and are thus strongly recommended for private use.

Weight and Balance

The maximum weight of the C206 varies from 3300lbs to 3600lbs, depending on model and modification. The most common configuration is 3600lbs.

Increased loading modifications are available, and many tip tank installations also provide an increase in maximum weight to 3800lbs. Normally a restriction is applied requiring the tip tanks to be a minimum of ½ full for the additional weight to be permitted due to wing loading. This restriction should be placarded close to the tip tank gauges of selector switch.

The unladen standard empty weight of the C206 is approximately 1800 to 2300lbs (820-1045kg) and includes full oil and unusable fuel. This provides a useful load of approximately 1500lbs.

The actual weight of the aircraft you are flying should always be used for weight and balance calculations. Refer to the relevant weight and balance certificate (which should be not older than 5 years and contained within in the aircraft's onboard documents) for exact weight of the aircraft you are flying.

Overloading, or incorrect loading, may not result in obvious structural damage, but can cause fatigue on internal structural components or produce hazardous aeroplane handling characteristics. An overweight or incorrectly loaded aircraft will have increased takeoff distance, climb rates, cruise speeds and landing distance and may become difficult to handle inflight.

Aeroplane balance is maintained by controlling the position of the Centre of Gravity. An aeroplane loaded past the rear limit of its permissible Centre of Gravity range will have an increased tendency for over-rotation, loss of elevator control on landing and, although a lower stall speed, a more unstable stall spin tendency. Aircraft loaded past the forward limit will result in a higher stall speed, and wheel-barrowing on takeoff or landing. An aircraft grossly out of balance in either direction will be uncontrollable.

If there is any significant weight in the rear compartment, load the largest passengers forward to help with the balance and with passenger comfort, as the rear seating area is very small, and if a cargo pod exists load the pod first. Always carefully check the loading, including the effect of fuel burn off during flight, which may have a significant affect on handling with a rear centre of gravity. If flying one or two up, without baggage, you may need to consider adding ballast to the aft compartment for better handling, particularly in the landing phase, depending on the position of the aircraft's centre of gravity and the aircraft model.

It sometimes may be necessary to calculate how far we can fly with the load on board then plan fuel stops in the required distance, in this case the planning calculations must simply be reversed, starting with the maximum weight and subtracting the load and aircraft empty weight to arrive at a fuel loading.

Weight and balance documentation is not normally required for private flights, however it is still the pilot in command's responsibility to ensure that the aircraft is properly loaded and within limits. It is vital for safety and performance considerations to know your exact operating weight and centre of gravity before each take-off, and the easiest and most fail safe way to confirm this is to complete a load sheet.

Do not listen to the 'hangar-talk' which very dangerously reports that a C206 can be loaded to it's volumetric capacity and flown safely as long as the tail does not touch the ground. This type of thinking has led to numerous take-off and landing accidents, but sadly the reports still exist.

Performance Graphs and Worksheets

Sample flight planning forms and some typical C206 performance tables and examples have been included for reference purposes. These tables are NOT for operational use, the Pilot's Operating Handbook in the aircraft you are flying MUST be consulted.

If there is ever any doubt in calculating performance, always use the figure which will provide the largest safety margin, or seek assistance.

Takeoff Performance

TAKEOFF DISTANCE
MAXIMUM WEIGHT 3600 LBS
SHORT FIELD

CONDITIONS:
Flaps 20°
2850 RPM, Full Throttle and Mixture Set at Placard Fuel Flow Prior to Brake Release
Cowl Flaps Open
Paved, Level, Dry Runway
Zero Wind

SAMPLE ONLY

MIXTURE SETTING	
PRESS ALT	GPH
S.L.	24
2000	23
4000	22
6000	21
8000	20

NOTES:
1. Short field technique as specified in Section 4.
2. Where distance value has been deleted, climb performance after lift-off is less than 150 fpm at takeoff speed.
3. Decrease distances 10% for each 10 knots headwind. For operation with tailwinds up to 10 knots, increase distances by 10% for each 2.5 knots.
4. For operation on a dry, grass runway, increase distances by 15% of the "ground roll" figure.

WEIGHT LBS	TAKEOFF SPEED KIAS		PRESS ALT FT	0°C		10°C		20°C		30°C		40°C	
	LIFT OFF	AT 50 FT		GRND ROLL	TOTAL TO CLEAR 50 FT OBS	GRND ROLL	TOTAL TO CLEAR 50 FT OBS	GRND ROLL	TOTAL TO CLEAR 50 FT OBS	GRND ROLL	TOTAL TO CLEAR 50 FT OBS	GRND ROLL	TOTAL TO CLEAR 50 FT OBS
3600	63	65	S.L.	810	1600	870	1715	935	1845	1000	1985	1075	2135
			1000	885	1755	950	1890	1020	2035	1095	2190	1175	2365
			2000	965	1935	1040	2085	1115	2250	1200	2430	1290	2630
			3000	1060	2140	1140	2310	1225	2500	1320	2710	1415	2945
			4000	1165	2380	1260	2575	1345	2795	1450	3040	1560	3320
			5000	1280	2660	1375	2890	1485	3145	1595	3445	1720	3790
			6000	1410	2995	1520	3270	1635	3580	1765	3950	1900	4390
			7000	1555	3405	1680	3740	1810	4135	1960	4615	---	---
			8000	1720	3925	1860	4360	2005	4890	---	---	---	---

Figure 5-4. Takeoff Distance (Sheet 1 of 2)

In the table above, for an aircraft at a pressure altitude of 2000ft and an expected temperature of 30 degrees at the time of departure, the ground roll is 1200ft and the total takeoff distance is 2430ft.

If the operation is conducted on dry grass, the factor should be increased by 15% of the ground roll, providing a figure of 2430+1200x0.15 = 2610ft.

If we apply a recommended safety figure of 1.33 to this to allow for differences in performance and conditions from those estimated, our total runway length required becomes 2610 x 1.33 = 3472ft or, dividing by 3.28, 1058m. This sounds like a high figure, but if you do not know the surface conditions or the particular aircraft's performance, and you don't want to end up in the trees, it is advisable to apply this safety factor.

In most older C206 manuals, figures are provided at maximum weight only. A lower weight would require a lower takeoff speed, however, a lower weight at the same speed will provide a shorter distance regardless. Since no figures are given for lower weights, the distances applicable for maximum weight must be used. If an emergency occurred during takeoff, for example an animal crossing the runway close to the point of rotation, or when applying a soft field technique, pilots should be aware that a lighter aircraft can get airborne at a lower speed.

Climb Performance

WEIGHT LBS	PRESS ALT FT	CLIMB SPEED KIAS	RATE OF CLIMB - FPM			
			-20°C	0°C	20°C	40°C
3600	S.L.	84	1080	990	895	805
	2000	83	940	855	770	680
	4000	82	810	725	640	555
	6000	81	680	595	515	430
	8000	79	550	470	390	310
	10,000	78	420	340	265	- - -
	12,000	77	295	220	145	- - -
3300	S.L.	82	1235	1145	1050	955
	2000	81	1090	1005	915	825
	4000	79	950	865	780	695
	6000	78	810	730	645	560
	8000	77	675	595	515	435
	10,000	76	540	460	385	- - -
	12,000	74	410	335	260	- - -
3000	S.L.	79	1420	1325	1225	1130
	2000	78	1260	1170	1080	985
	4000	77	1110	1025	940	855
	6000	76	965	880	800	715
	8000	74	815	740	655	575
	10,000	73	675	595	520	- - -
	12,000	71	535	460	385	- - -

Using the same example, an aircraft taking off at 2000ft at maximum weight and with an ambient temperature of 30 degrees, and considering the first 1500 feet of climb, the expected climb performance would be somewhere between 555 and 770 feet per minute. Of course it should not be as low as 555 or as high as 770, and we could interpolate, but this is not of much practical use, since the actual aircraft performance cannot be predicted so precisely. These figures are close enough for fuel calculation purposes, and for safety and simplicity it is better to round down to 500fpm.

The best rate of climb speed is 83kts at 2000ft, the speed reducing as the aircraft climbs by approximately one knot per 2000ft. Again this estimate is on the higher side and therefore safer side.

Cruise Performance

Cruise performance tables provide accurate fuel flow and true air speed at a variety of altitudes and power settings.

For this example we have selected 6000ft as the cruise altitudes, and the ambient temperature on the ground at 2000ft was 30 degrees. On the following page the cruise table for a pressure altitude of 6000ft can be seen. If 6000ft happens to be below the transition altitude, the difference between altitude and pressure altitude is insignificant for these calculations and need not be considered.

To determine estimated cruise performance, we first need to calculate the estimated temperature at the planned cruise altitude. The table allows for ISA-20, ISA and ISA+20, and the fuel flows are considerably different for each. (Note the different application of temperature from the previous graphs, this can be confusing if table headings are not read properly, and will vary between models).

Using a standard lapse rate of 2 degrees per thousand feet, the temperature at 6000ft should be approximately 30-2x4 (4000ft change in altitude) = 22 degrees.

The standard temperature at 2000ft is 15-2x2 = 11 degrees, so at the airfield the deviation is 19 degrees above standard, which would be the same at altitude since we have estimated using the standard lapse rate. If a forecast upper level temperature was available, it should be used, as lapse rates are not always standard, in this case, and if a temperature is not given on the table, the calculation would need to be applied to the cruising level.

Here the table provides a temperature of 23 degrees as 20 degrees above standard at 6000ft, providing a cross check on our calculation.

We can use the figures in the far right column, again rounding the temperature up is erring on the safe side for performance.

Selecting cruise power of 2400rpm and 23" of manifold pressure will result in an estimated true air speed of 139kts, and a fuel flow of 13.5 US gallons per hour (x 3.785 =51 litres). Applying a contingency figure of 10% provides a figure of 139/1.1 = 125kts and 13.5x1.1 = 15.3gph, or 58 litres, which is much closer to the block figures most C206 commercial operators recommend of 120kts and 60lts per hour. This illustrates the importance of safety factors, especially when operating a particular model or tail number for the first time.

Cruise Performance Tables

CRUISE PERFORMANCE
PRESSURE ALTITUDE 6000 FEET

SAMPLE ONLY

CONDITIONS:
3600 Pounds
Recommended Lean Mixture
Cowl Flaps Closed

NOTE
For best fuel economy at 65% power or less, operate at 1 GPH leaner than shown in this chart or at peak EGT if an EGT indicator is installed.

RPM	MP	20°C BELOW STANDARD TEMP -17°C			STANDARD TEMPERATURE 3°C			20°C ABOVE STANDARD TEMP 23°C		
		% BHP	KTAS	GPH	% BHP	KTAS	GPH	% BHP	KTAS	GPH
2550	24	---	---	---	78	148	16.2	75	149	15.7
	23	76	144	16.0	74	145	15.4	71	145	14.9
	22	72	141	15.1	69	141	14.5	67	142	14.1
	21	68	137	14.2	65	137	13.7	63	138	13.3
2500	24	78	145	16.3	75	146	15.8	73	147	15.2
	23	74	142	15.5	71	143	14.9	69	143	14.4
	22	70	139	14.6	67	139	14.1	65	140	13.7
	21	66	135	13.8	63	135	13.3	61	136	12.9
2400	24	73	141	15.2	70	142	14.7	68	142	14.2
	23	69	138	14.5	67	138	14.0	64	139	13.5
	22	65	134	13.7	63	135	13.2	61	135	12.8
	21	61	131	12.9	59	131	12.5	57	131	12.1
2300	24	68	137	14.3	66	138	13.8	64	138	13.4
	23	65	134	13.6	62	135	13.1	60	135	12.7
	22	61	130	12.9	59	131	12.4	57	131	12.1
	21	57	127	12.1	55	127	11.8	53	127	11.4
2200	24	63	132	13.3	61	133	12.8	59	133	12.4
	23	60	129	12.6	58	130	12.2	56	130	11.8
	22	57	126	12.0	54	126	11.6	53	126	11.2
	21	53	122	11.3	51	122	11.0	49	122	10.7
	20	50	118	10.7	48	118	10.3	46	117	10.0
	19	46	113	10.0	44	113	9.7	43	112	9.4

Landing Distance

LANDING DISTANCE
SHORT FIELD

CONDITIONS:
Flaps 40°
Power Off
Maximum Braking
Paved, Level, Dry Runway
Zero Wind

SAMPLE ONLY

NOTES:
1. Short field technique as specified in Section 4.
2. Decrease distances 10% for each 10 knots headwind. For operation with tailwinds up to 10 knots, increase distances by 10% for each 2.5 knots.
3. For operation on a dry, grass runway, increase distances by 40% of the "ground roll" figure.

WEIGHT LBS	SPEED AT 50 FT KIAS	PRESS ALT FT	0°C		10°C		20°C		30°C		40°C	
			GRND ROLL	TOTAL TO CLEAR 50 FT OBS	GRND ROLL	TOTAL TO CLEAR 50 FT OBS	GRND ROLL	TOTAL TO CLEAR 50 FT OBS	GRND ROLL	TOTAL TO CLEAR 50 FT OBS	GRND ROLL	TOTAL TO CLEAR 50 FT OBS
3600	64	S.L.	695	1340	720	1375	750	1415	775	1450	800	1490
		1000	720	1375	750	1415	775	1450	800	1490	830	1530
		2000	750	1415	775	1455	805	1495	830	1530	860	1575
		3000	775	1455	805	1495	835	1540	865	1580	890	1615
		4000	805	1495	835	1540	865	1580	895	1625	925	1665
		5000	835	1540	870	1585	900	1630	930	1675	960	1715
		6000	870	1590	900	1630	935	1680	965	1725	995	1770
		7000	905	1635	935	1680	970	1730	1000	1775	1035	1825
		8000	940	1690	970	1730	1005	1780	1040	1830	1075	1880

Landing back at the same airfield, and with the same conditions, 30 degrees Celsius and 2000ft pressure altitude, we obtain a figure of 830ft for the ground roll and 1530ft for the total distance, however this time the factor for landing on dry grass is 40%. This provides a total distance of 0.4x830+1530 = 1862ft. Applying a factor of 1.43, a higher figure is recommended for landing due to the fact that technique can play a much bigger part in the performance, gives a total landing distance of 1.43x1862 = 2663ft or 812m.

The figures in this table illustrate truth about the great landing performance the C206 is known for, but beware, the takeoff will nearly always be more limiting.

Landing performance again is only provided in this model's handbook for maximum weight, and the same theory applies for takeoff, however pilots should be aware that the further the speed varies from the minimum margin permitted above the stall, 1.3vso, the more the aircraft will tend to float during the flare and take longer to reduce to the required touch down speed.

Non-manufacturer Performance Factors

Most Cessna flight manuals only provide figures for paved surfaces with a degradation factor for dry grass.

Common situations which are not provided for include wet grass, soft or rough ground and snow. All of these have severe performance penalties, none are prohibited, yet the manufacturer gives us no guidance on how to cater for them.
Because of this, some civil aviation authorities have provided recommended performance factors for applying where no other guidance is available. This allows a safe estimate of performance, which is far preferable to the alternative method of a pilot's educated guess.

The table provided on the following page is reproduced from the UKCAA safety sense booklets provided free on the UKCAA website. It is a guideline for recommended safety factors to apply to performance figures where the aircraft's flight manual does not specify a factor.

When using performance factors, remember each factor must be multiplied. Soft ground with wet grass, a common situation in many countries, would attract a factor of 1.25 x 1.3 = 1.625. Adding the additional safety factor of 1.33 to this would increase by 1.62 x 1.33 = 2.15.

Figures applied to surface conditions naturally are only applied to the ground roll, which is typically around 50-70% the total distance, so in this example the entire length would increase by approximately 1.5 to 1.7 times the standard paved surface figure.

Whenever the exact nature of the surface is unfamiliar, application of the safety factor (1.33 for takeoff and 1.43 for landing), which is compulsory for most commercial operations under all conditions, is strongly recommended.

Getting to know your aircraft performance in these situations on a runway where length is more than adequate can permit reductions in non mandatory factors, but never attempt to operate in conditions which are less than the figures provided the aircraft's manual and never underestimate the effects of changes in weight, surface conditions, and density on your performance at a field that you are unfamiliar with.

The performance sheets provided in this book have a section for applying factors to both ground roll or total distance.

Non-manufacturer Performance Factors

FACTORS MUST BE MULTIPLIED i.e. 1·2 x 1·3

CONDITION	TAKE-OFF		LANDING	
	INCREASE IN DISTANCE TO HEIGHT 50 FEET	FACTOR	INCREASE IN LANDING DISTANCE FROM 50 FEET	FACTOR
A 10% increase in aeroplane weight, e.g. another passenger	20%	1.2	10%	1.1
An increase of 1,000 ft in aerodrome elevation	10%	1.1	5%	1.05
An increase of 10°C in ambient temperature	10%	1.1	5%	1.05
Dry grass* – Up to 20 cm (8 in) (on firm soil)	20%	1.2	20% +	1.2
Wet grass* – Up to 20 cm (8 in) (on firm soil)	30%	1.3	30% + 1.3 When the grass is very short, the surface may be slippery and distances may increase by up to 60%.	
A 2% slope*	uphill 10%	1.1	downhill 10%	1.1
A tailwind component of 10% of lift-off speed	20%	1.2	20%	1.2
Soft ground or snow*	25% or more	1.25	25% + or more	1.25
NOW USE ADDITIONAL SAFETY FACTORS (if data is unfactored)		1.33		1.43

Notes: 1. * Effect on Ground Run/Roll will be greater.
2. + For a few types of aeroplane e.g. those without brakes, grass surfaces may decrease the landing roll. However, to be on the safe side, assume the INCREASE shown until you are thoroughly conversant with the aeroplane type.
3. Any deviation from normal operating techniques is likely to result in an increased distance.

So, if the distance required exceeds the distance available, changes will HAVE to be made.

Ground Planning Worksheets and In-flight Logs

All the following performance sheets are provided for operational or training use, wherever operator documents are not available. Hard copies may be found at http://www.redskyventures.org, under the Free Stuff tab.

Navigation Calculation Work Sheet

Date: / / REG:											

FM	TO	FL	Temp	W/V	IAS	TAS	DRIF T	Hdg T	VAR.	Hdg M	G/S	Dist	EET
TOTALS													

Fuel Planning Work Sheet

	LITRES
1. ENROUTE TIME @ _____ LITRES / HOUR	
2. TAXI / TAKEOFF / APPROACH / LANDING	
TOTAL TRIP FUEL	
3. 10 % CONTINGENCY FUEL	
4. RESERVE (45 MINS) @ _____ LITRES / HOUR	
5. UNUSABLE FUEL	
MIN FUEL REQUIRED	
6. ADDITIONAL FUEL (PIC discretion)	
MIN REQUIRED FUEL	
TOTAL FUEL DIPPED	
LESS UNUSABLE FUEL (Included in aircraft empty weight)	-
LITRES TO POUNDS (AVGAS 100LL)	x 1.584
TOTAL FUEL WEIGHT (TO WEIGHT AND BALANCE SHEET)	
TOTAL FUEL BURN (1 +2)	

Weight and Balance Work Sheet

ITEM	WEIGHT	ARM	MOMENT / 1000
Aircraft Empty Weight (Flt. Docs)			
Pilot, front passenger			
Centre Passenger(s)			
Rear Passenger(s)			
Baggage Area 1 (Max _____ lbs)			
Baggage Area 2 (Max _____ lbs)			
Fuel Weight 1 (Max _____ lbs)			
Fuel Weight 2 (Max _____ lbs)			
Other_____			
Other_____			
Initial Takeoff Weight (Max _____ lbs)			
+/- Additional FUEL Adjustment			
FINAL TAKEOFF WEIGHT (Max _____ lbs)			
Less Fuel Burn			
LANDING WEIGHT (Max _____ lbs)			

Weight x Arm = Moment.
Total Moment = Sum of all Moments (+ or -)
Total Weight = Sum of all Weights (+ or -)
Final C. of G. = Total moment / Total weight

CESSNA 206 TRAINING MANUAL

Departure Performance Work Sheet

DEPARTURE AIRFIELD:			DATE:			(dd-mmm-yy)
PIC:			AIRCRAFT:			REG:

NOTE: ALL Calculations require correct integer (+ or – sign) to be carried through

(1) Pressure altitude (PA) = Altitude AMSL + 30 x (1013-QNH)

Standard QNH	Minus Airfield QNH	Equals (+/-)	ft per mb	Equals (+/-)	+ELEVATION	**PRESSURE ALTITUDE**
1013	-		x30			

(2) Standard Temperature ST=15–2xPA/1000 ie. 2 degrees Celsius cooler per 1000ft altitude (Use only if not allowed for on the performance graphs)

Pressure ALT	Divide by 1000	Equals	Multiply by (-2) deg per deg Celsius	Equals (-)	Add 15	**STANDARD TEMP**
	/1000		x-2		+15C	

(3) Density altitude (DA) DA = PA +(-) 120ft/deg above (below) ST
(Use only if not allowed for on Graphs)

+/-ACTUAL TEMP	minus +/- STD TEMP	Equals (+/-)	Multiplied by ft per degree	Equals (+/-)	+Press Alt	**DENSITY ALTITUDE**
			x120			

Wind degrees True	Deviation +W/-E	Wind Mag	Runway Heading	Magnetic Difference	Multiply by Closest Factor	Wind In Knots	**Approx. HWC/XWC**
				X-	30=x0.5		XWC-
				H-	45=x0.7		HWC-
					60=x0.9		
				T-full	T = 1.0		TWC -

Surface	Dry/Wet/Paved/Grass/Gravel/Other_____	Slope:	U
			P

TAKE OFF ROLL REQUIRED	
FACTORS FOR GROUND ROLL_____	
BASIC TAKEOFF DISTANCE	
FACTORS: WIND_____ SLOPE_____ SURFACE_____ SAFETY 1.33__ OTHER_____	TOTAL FACTOR _____
TOTAL RUNWAY LENGTH REQUIRED	
TAKEOFF DISTANCE AVAILABLE	

Arrival Performance Work Sheet

ARRIVAL AIRFIELD:		DATE:	(dd-mmm-yy)
PIC:		AIRCRAFT:	REG:

NOTE: ALL Calculations require correct integer (+ or − sign) to be carried through

(1) Pressure altitude (PA) = Altitude AMSL + 30 x (1013−QNH)

Standard QNH	Minus Airfield QNH	Equals (+/−)	ft per mb	Equals (+/−)	+ELEVATION	**PRESSURE ALTITUDE**
1013	−		x30			

(2) Standard Temperature ST=15−2xPA/1000 ie. 2 degrees Celsius cooler per 1000ft altitude (Use only if not allowed for on Graphs)

Pressure ALT	Divide by 1000	Equals	Multiply by (−2) deg per deg Celsius	Equals (−)	Add 15	**STANDARD TEMP**
	/1000		x−2		+15C	

(3) Density altitude (DA) DA = PA +(−) 120ft/deg above (below) ST
(Use only if not allowed for on Graphs)

+/−ACTUAL TEMP	minus +/− STD TEMP	Equals (+/−)	Multiplied by ft per degree	Equals (+/−)	+Press Alt	**DENSITY ALTITUDE**
			x120			

Wind degrees True							
Wind degrees True	Deviation +W/−E	Wind Mag	Runway Heading	Magnetic Difference	Multiply by Closest Factor	Wind in Knots	**Approx. HWC/XWC**
				X−	30=x0.5		XWC−
				H−	45=x0.7		HWC−
				T-full	60=x0.9		TWC − (full)
					T = 1.0		

Surface Dry/Wet/Paved/Grass/Gravel/Other_____   Slope: D
                                                             N

LANDING GROUND ROLL REQUIRED	
FACTORS FOR GROUND ROLL_____	
TOTAL LANDING DISTANCE REQUIRED	

FACTORS: WIND_____ SLOPE_____ SURFACE_____ TOTAL FACTOR _____
SAFETY 1.43   OTHER_____

TOTAL RUNWAY LENGTH REQUIRED
LANDING DISTANCE AVAILABLE

Navigation Log

FM	TO	FL	TRK True	W/V	HDG True	HDG Mag	Dist	G/S	EET	ETA1	ETA2	ETA3	ATA
TOTALS													

LEFT TANK	Start Fuel:			RIGHT TANK	Start Fuel:		
TIME ON	FUEL USED	REMAINING		TIME ON	FUEL USED	REMAINING	

CLEARANCES/NOTES			WEATHER			
			STN			
			CODE			
			RWY			
			W/V			
			Vis			
			Cld			
			Cld			
			WX			
			WX			
			Temp			
			QNH			
			Other			
			Other			

REVIEW QUESTIONS

Engine and Engine Systems

1. If, while the engine is running, the magneto selector is inadvertently turned to the OFF position:
   a) there will be a drop in engine rpm;
   b) the rpm will not be affected;
   c) the engine will stop.

2. Two separate ignition systems provide:
   a) more safety only;
   b) more efficient burning only;
   c) more safety and more efficient burning;
   d) dual position key switching.

3. Switching the ignition OFF connects the magneto system to ground:
   a) true;
   b) false.

4. If a magneto ground wire comes loose in flight, the engine:
   a) will stop;
   b) will continue running with lower rpm;
   c) will continue running with no change.

5. The spark plugs are provided with an electrical supply from:
   a) the battery at all times;
   b) the magnetos at all times;
   c) the battery at start-up and then the magnetos.

6. The most probable reason an engine continues to run after ignition switch has been turned off is:
   a) carbon deposit glowing on the spark plugs;
   b) a magneto ground wire is in contact with the engine casing;
   c) a broken magneto ground wire.

7. The maximum drop and maximum differential on the magneto check for the C206 is:
   a) 150 drop, 75 difference;
   b) 50 drop, 150 difference;
   c) 125 drop, 75 difference.

8. Cessna 206 engine has:
   a) fuel injection;
   b) carburettor
   c) either fuel injection or carburettor depending on the model.

9. Cessna 206 engines are:
   a) all models are sensitive to carburettor ice;
   b) not affected by carburettor ice, since it does not have a carurettor;
   c) sensitive to carburettor ice when equipped with a carburettor.

10. During a magneto check a drop of 300rpm is experienced on the left magneto, and a drop of 100rpm on the right magneto, but in both positions the engine is running smoothly, the most likely explanation is:
    a) spark plug fouling;
    b) a faulty left magneto;
    c) there is nothing wrong, the engine is running normally.

11. The pilot controls the fuel/air ratio with the:
    a) throttle;
    b) carb. Heat;
    c) mixture;
    d) the engine is self adjusting and there is no control for the fuel/air ratio.

12. For takeoff at a sea level airport, the mixture control should be:
    a) in the leaned position for maximum rpm;
    b) in the full rich position;
    c) it does not matter since the engine is not affected by mixture setting below 3000ft.

13. What will occur if the mixture control is left full rich as the flight altitude increases:
    a) the volume of air entering the cylinder increases while the amount of fuel decreases, resulting in a lean mixture;
    b) the density of air entering the cylinder decreases while the amount of fuel increases, resulting in a rich mixture;
    c) the density of air entering the cylinder decreases while the amount of fuel remains constant, resulting in a rich mixture.

14. The correct procedure to achieve the best fuel/air mixture when cruising at altitude is:
    a) to move the mixture control toward LEAN until engine rpm starts to drop;
    b) to move the mixture control toward LEAN until engine rpm reaches a peak value;
    c) to move the mixture control toward RICH until engine rpm starts to drop;
    d) to move the mixture control toward LEAN until engine rpm reaches a peak EGT or rpm and then toward RICH to get EGT 25-50°F below the peak, approximately 3 turns.

15. The extra fuel in a rich mixture causes:
    a) engine heating;
    b) engine cooling;
    c) does not affect the heating or cooling of the engine.

16. If, after the mixture is properly adjusted during cruise at the altitude, the pilot forgets to enrich the mixture during descent:
    a) a too rich mixture will assist with cooling, but the engine may foul plugs or cut-out because the mixture is too rich;
    b) a to lean mixture will create high cylinder head temperatures, and the engine may cut-out because the mixture is too lean;
    c) the descent will have no effect on the fuel-air mixture ratio.

CESSNA 206   TRAINING MANUAL

17. The engine oil system is provided to:
    a) reduce friction between moving parts and ensure high engine temperatures;
    b) reduce friction between moving parts and prevent high engine temperatures;
    c) increase friction between moving parts and prevent high engine temperatures.

18. Oil grades:
    a) should not be mixed;
    b) may be mixed;
    c) may be mixed only when there is no viable alternative.

19. With too little oil, you may observe:
    a) high oil temperature and high oil pressure;
    b) high oil temperature and low oil pressure;
    c) low oil temperature and low oil pressure.

20. What action can a pilot take to aid in cooling an engine that is overheating during a climb:
    a) lean the mixture and increase airspeed;
    b) enrichen the mixture and increase airspeed;
    c) reduce engine rpm and open the cowl flaps;
    d) Both a and c.

21. The pilot should shut-down an engine after start if the oil pressure does not rise within:
    a) 30 seconds;
    b) 1 minutes;
    c) 10 seconds.

22. The aircraft is equipped with:
    a) a fixed pitch propeller;
    b) a constant speed propeller.

23. Engine power is monitored by the:
    a) manifold pressure gauge;
    b) engine rpm gauge;
    c) both the manifold pressure and rpm depending on the condition of the propeller.

24. The usual method of shutting an engine down is to:
    a) switch the magnetos off;
    b) move the mixture to idle cut-off;
    c) switch the battery master switch off.

25. The minimum oil quantity for start and normal flight is:
    a) 12 quarts;
    b) 7 quarts;
    c) 9 quarts.

26. Cowl flaps should be open:
   a) at all times;
   b) for climb and ground operations and whenever high engine operating temperatures are experienced;
   c) whenever the engine temperature is very low.

27. Cowl flaps should be closed:
   a) at all times;
   b) for climb and ground operations and high engine operating temperatures;
   c) whenever the engine operating temperature is too low.

Fuel System

28. Fuel tanks is are located:
   a) in the aft cabin;
   b) beneath the pilot seats;
   c) in the wings.

29. Water tends to collect at the:
   a) lowest point in the fuel system;
   b) highest point in the fuel system.

30. The average consumption rate of the C206 is approximately
   a) 65 liters;
   b) 75 liters;
   c) 55 liters.

31. The fuel cock selections are:
   a) both only;
   b) left, right and both, left and right can be used for level flight only;
   c) left, right and both and all positions can be used in any phase of flight.

32. The fuel pump is used for:
   a) priming and engine driven pump failure only;
   b) priming, takeoff and landing and engine driven pump failures;
   c) priming, purging, engine driven pump failure and vapor surges.

33. The fuel pump should be selected to:
   a) HI for priming, HI or LOW depending on demand for engine driven pump failure and vapor surges;
   b) LOW for priming, HI for engine driven pump failure and vapor surges;
   c) HI or LOW depending on demand for engine driven pump failure, priming and vapor surges.

Airframe, Electrical and Instruments

34. Normal in-flight electrical power is provided by an:
   a) alternator;
   b) battery;
   c) generator.

35. A distribution point for electrical power to various services is:
    a) circuit breaker;
    b) distributor;
    c) bus bar.

36. The battery master switch should be turned to OFF after the engine is stopped to avoid the battery discharging through:
    a) the magnetos;
    b) the generator;
    c) electrical services connected to it.

37. The suction (or vacuum gauge) shows the pressure:
    a) below atmospheric pressure;
    b) above atmospheric pressure;
    c) This gauge does not read units of pressure.

38. The vacuum pump is:
    a) electrically-driven;
    b) engine-driven;
    c) hydraulically-driven.

39. The following instrument will be affected by a vacuum pump failure:
    a) artificial horizon and direction indicator;
    b) turn and bank indicator;
    c) airspeed indicator.

40. The aircraft is equipped with:
    a) electrically operated elevator trim tab;
    b) manually-operated elevator trim;
    c) manually-operated elevator and rudder trim;
    d) manual rudder trim and electric or manual elevator trim depending on the options installed.

41. Frise type ailerons are used to:
    a) reduce airflow over the control surface to make the control lighter;
    b) reduce the adverse aileron yaw during bank;
    c) this aircraft does not have Frise type of ailerons;

42. The flaps are:
    a) hydraulically-operated;
    b) electrically-operated;
    c) may be electric or manual depending on model.

43. Flaps selections are:
    a) 10, 20 and 40 degrees;
    b) take-off, approach and land;
    c) 10, 20 and 30 degrees.

44. Nose wheel steering is provided by:
   a) mechanical links with rudder pedals;
   b) differential braking;
   c) all of the above.

Flight Operations

Fill in the following from the aircraft you are flying, model _____, year _____.

45. The best rate of climb speed at sea level is _____, at 10'000ft _____.

46. The best angle of climb speed at sea level is _____, at 10'000ft _____.

47. The recommended takeoff speed at sea level, and maximum weight for a short field is _____, and for a normal landing is _____)_.

48. The recommended landing speed at sea level and maximum weight for a short field is _____, and for a normal landing is _____.

49. The best glide speed at maximum weight is _____.

50. The recommended speed for an engine failure after takeoff is _____.

www.ingramcontent.com/pod-product-compliance
Lightning Source LLC
Chambersburg PA
CBHW071426180526
45170CB00001B/240